SpringerBriefs in Earth System Sciences

Series editors

Gerrit Lohmann, Universität Bremen, Bremen, Germany

Lawrence A. Mysak, Department of Atmospheric and Oceanic Sciences, McGill University, Montreal, QC, Canada

Justus Notholt, Institute of Environmental Physics, University of Bremen, Bremen, Germany

Jorge Rabassa, Laboratorio de Geomorfología y Cuaternario, CADIC-CONICET, Ushuaia, Tierra del Fuego, Argentina

Vikram Unnithan, Department of Earth and Space Sciences, Jacobs University Bremen, Bremen, Germany

W0192935

SpringerBriefs in Earth System Sciences present concise summaries of cutting-edge research and practical applications. The series focuses on interdisciplinary research linking the lithosphere, atmosphere, biosphere, cryosphere, and hydrosphere building the system earth. It publishes peer-reviewed monographs under the editorial supervision of an international advisory board with the aim to publish 8 to 12 weeks after acceptance. Featuring compact volumes of 50 to 125 pages (approx. 20,000–70,000 words), the series covers a range of content from professional to academic such as:

- A timely reports of state-of-the art analytical techniques
- bridges between new research results
- snapshots of hot and/or emerging topics
- literature reviews
- in-depth case studies

Briefs are published as part of Springer's eBook collection, with millions of users worldwide. In addition, Briefs are available for individual print and electronic purchase.

Briefs are characterized by fast, global electronic dissemination, standard publishing contracts, easy-to-use manuscript preparation and formatting guidelines, and expedited production schedules.

Both solicited and unsolicited manuscripts are considered for publication in this series.

More information about this series at http://www.springer.com/series/10032

Jack J. Middelburg

Marine Carbon Biogeochemistry

A Primer for Earth System Scientists

 Springer Open

Jack J. Middelburg
Department of Earth Sciences
Utrecht University
Utrecht, The Netherlands

ISSN 2191-589X   ISSN 2191-5903   (electronic)
SpringerBriefs in Earth System Sciences
ISBN 978-3-030-10821-2   ISBN 978-3-030-10822-9   (eBook)
https://doi.org/10.1007/978-3-030-10822-9

Library of Congress Control Number: 2018965889

This Springer imprint is published by the registered company Springer Nature Switzerland AG
The registered company address is: Gewerbestrasse 11, 6330 Cham, Switzerland

Preface

Biogeochemistry, a branch of Earth System Sciences, focusses on the two-way interactions between organisms and their environment, including the cycling of energy and elements and the functioning of organisms and ecosystems. To this end, physical, chemical, biological and geological processes are studied using field observations, experiments, modelling and theory. The discipline of biogeochemistry has grown to such an extent that sub-disciplines have emerged. Consequently, producing a single comprehensive textbook covering all aspects, e.g., terrestrial, freshwater and marine domains, biogeochemical cycles and budgets of the major biological relevant elements, reconstruction of biogeochemical cycles in the past, earth system modelling, microbiological, organic and inorganic geochemical methods, theory and models, has become unworkable.

This book provides a concise treatment of the main concepts in ocean carbon cycling research. It focusses on marine biogeochemical processes impacting the cycling of particulate carbon, in particular organic carbon. Other biogeochemical processes impacting nitrogen, phosphorus, sulphur, etc., and the identity of the organisms involved are only covered where needed to understand carbon biogeochemistry. Moreover, chemical and biological processes relevant to carbon cycling are central, i.e. for physical processes, the reader might consult the excellent ocean biogeochemical dynamics textbooks of Sarmiento and Gruber (2006; Princeton University Press) and Williams and Follows (2011; Cambridge University Press). My text aims to provide graduate students in marine and earth sciences a conceptual understanding of ocean carbon biogeochemistry, so that they are better equipped to read palaeorecords, can improve carbon biogeochemical models and generate more accurate projections of the functioning of the future ocean. Because the book is targeted at students having a background in environmental and earth sciences, some basic biological concepts are explained. Some basic understanding of calculus is expected. Simple mathematical models are used to highlight the most important factors governing carbon cycling in the ocean. The material here is based on a selection of lectures in my Utrecht University master course on Microbes and Biogeochemical Cycles.

This first draft of this book was written during a three-month sabbatical stay at Department of Geosciences, Princeton University (April–June 2018). I thank Bess Ward, chair of that department, for providing a desk and a stimulating environment.

This sabbatical stay was supported by a travel grant from the Netherlands Earth System Science Centre. I thank Bernie Boudreau for carefully scrutinizing the initial draft, Mathilde Hagens and Karline Soetaert for feedback on Chap. 5 and Anna de Kluijver for remarks on Chap. 6. Ton Markus improved my draft figures. Finally, I thank my wife and publisher Petra van Steenbergen.

Utrecht, The Netherlands   Jack J. Middelburg

Contents

Symbols

B Biomass of phytoplankton or buffer value (Chap. 5)

D Diffusion coefficient (area time^{-1}); with D_s: diffusion of solutes in sediments, D_b: particle mixing in sediments

E Radiant energy (mol quanta area^{-1} time^{-1})

E_A Activation energy (J mol^{-1})

F Flux of material (mol/gr area^{-1} time^{-1})

G Quantity of organic carbon (mol/gr C per gr sediment, area or volume)

k First-order rate/decay constant (time^{-1})

k_{PAR} Light extinction coefficient (length^{-1})

K Half-saturation constant in Monod-type equation; K_E: light saturation parameter; K_μ: growth (nutrient) half-saturation constant

K_x Equilibrium constants (x = w, H, 1, 2) that depend on temperature, pressure and solution composition

K_z Eddy-diffusion (mixing) coefficient in water column (area time^{-1})

P Production (mol/gr volume^{-1} time^{-1})

Q_{10} Increase in rate for 10 °C increase in T

Q Cellular quota in Droop equation

r First-order rate constant for phytoplankton (time^{-1})

R_0 Zero-order production or consumption term (mol/gr volume^{-1} time^{-1})

t Time

T Temperature (°C or K)

w Particle settling in water column or sediment accumulation rate (length time^{-1})

x Depth in sediment (length)

z Depth in water column (length)

z_{eu} Euphotic zone depth (length)

z_c Compensation depth (length) where phytoplankton growth and respiration are equal

z_{cr} Critical depth (length) where phytoplankton production balances losses

β Buffer value in terms of proton concentration

β_{tr} Solute transfer coefficient at seafloor

ϕ Porosity in sediment
μ Maximum growth rate (time^{-1})
θ Maximum nutrient uptake
ρ Dry density of particle (gr volume^{-1})
ψ Carbon dioxide generated per unit carbonate precipitated

Introduction

1

The name biogeochemistry implies that it is a discipline integrating data, knowledge, concepts and theory from biology, geosciences and chemistry. Biogeochemists extensively use approaches from a wide range of disciplines, including physical, chemical and biological oceanography, limnology, atmospheric sciences, ecology and microbiology, civil and environmental engineering, soil science and geochemistry. This diversity in scientific backgrounds stimulates cross-fertilization and research creativity, which are needed to elucidate the reciprocal relationships between living organisms and their environment at multiple scales during times of global change. Biogeochemistry aims to provide a holistic picture of natural ecosystem functioning. The challenge is to identify the right level of detail needed to understand the dynamics of elemental cycles and the functioning of biological communities. This implies that single-cell organism level studies and molecular orbital calculations of chemical reactions require upscaling to the appropriate temporal and spatial scale (often involving first-principle physics based models) to understand how natural ecosystems deal with perturbations and how life has shaped our planet.

Although biogeochemistry developed as a full discipline in the mid-1980s with the launch of the international geosphere-biosphere program (IGBP, 1987) and the journals Biogeochemistry (1984) and Global Biogeochemical Cycles (1987), its roots can be traced back to early scientists documenting how living organisms transformed chemical substances, such as oxygen production during photosynthesis (Priestly, 1733–1804), phosphorus in organisms' tissues (Lavoisier, 1743–1789) and nitrogen fixation by bacteria (Beijerinck, 1851–1931). Naturalist and avant-la-lettre multidisciplinary scientists, such as Alexander von Humboldt (1769–1859). Charles Darwin (1808–1882) and Alfred Lotka (1880–1949), pioneered what we would recognize as biogeosciences in the 21st century. Darwins' studies of atmospheric deposition, bioturbation and formation and sustenance of coral reefs are still key areas in modern biogeochemistry. The tight relationship between living organisms and their environment figured prominently in Lotka's book "Elements of Physical Biology" (1925): "*It is not so much the organism or the species that*

© The Author(s) 2019
J. J. Middelburg, *Marine Carbon Biogeochemistry*, SpringerBriefs in Earth
System Sciences, https://doi.org/10.1007/978-3-030-10822-9_1

evolves, but the entire system, species and environment. The two are inseparable."
This concept that organisms shape the environment and govern elemental cycles on
Earth underlies the biosphere concept of Vladimir Vernadsky (1863–1945), a
geochemist and mineralogist, often considered the founder of biogeochemistry.
G. Evelyn Hutchinson (1903–1991) was instrumental in establishing biogeochemical, whole-system approaches to study lakes. Alfred Redfield (1890–1983)
discovered that nitrogen to phosphorus ratios of phytoplankton in seawater are
constant and similar to dissolved ratios, implying co-evolution of the environment
and organisms living in it. His seminal 1958 article started as follows "*It is a
recognized principle of ecology that the interaction of organisms and environment
are reciprocal. The environment not only determines the conditions under which
life exists, but the organisms influence the conditions prevailing in their environment*" (Redfield 1958). The latter was articulated in the Gaia hypothesis of Lovelock (1972): The Earth became and is maintained habitable because of multiple
feedback mechanisms involving organisms. For instance, biologically mediated
weathering of rocks removes carbon dioxide from the atmosphere and generates
bicarbonate and cations that eventually arrive in the ocean, where calcifiers produce
the minerals calcite and aragonite and release carbon dioxide back to the
atmosphere.

The above one-paragraph summary of the history of biogeochemistry does not
mean that it was a linear or smooth process. While the early pioneers (before the
second world war) were not hindered much by disciplinary boundaries between
physics, biology, chemistry and earth sciences, the exponential growth of scientific
knowledge and the consequent specialization and success of reductionism to
advance science, had led to an under appreciation of holistic approaches crossing
disciplinary boundaries during the period 1945–1990. Addressing holistic research
questions may require development of new concepts and methods, but often
involves application and combination of well-established theory or methods from
multiple disciplines. The latter implies finding the optimal balance between biology,
chemistry and physics to advance our understanding of biogeochemical processes.
For instance, all biogeochemical models have to trade-off spatial resolution in the
physical domain with the number of chemical elements/compounds and the
diversity of organisms to be included. Ignoring spatial dimensions and heterogeneity through the use of box models may seem highly simplistic to a physical
oceanographer, but may be sufficient to obtain first-order understanding of elemental cycling. Similarly, organic carbon flows can be investigated via study of the
organisms involved, the composition of the organic matter or by quantifying the
rates of transformation, without considering the identity of the organisms involved.
Each disciplinary approach has its strengths and weaknesses, and they are unfortunately not always internally consistent. However, this confrontation of different
disciplinary concepts has advanced our understanding (Middelburg 2018). In the
next section, we will discuss why many geochemists embraced biogeochemistry.

1.1   From Geochemistry and Microbial Ecology to Biogeochemistry

Geochemistry is a branch of earth sciences that applies chemical tools and theory to study earth materials (minerals, rocks, sediments and water) to advance understanding of the Earth and its components. While early studies focused on the distribution of elements and minerals using tools from analytical chemistry, the next step involved the use of chemical thermodynamics to explain and predict the occurrence and assemblages of minerals in sediments and rocks. The thermodynamic approach was and is very powerful in high-temperature systems (igneous rocks, volcanism, metamorphism, hydrothermal vents), but it was less successful in predicting geochemical processes at the earth surface. Geochemists studying earth surface processes soon realized that predictions based on thermodynamics, i.e. the Gibbs free energy change of a reaction, provided a necessary condition whether a certain reaction could take place, but not a sufficient constraint whether it would take place because of kinetics and biology.

Realizing the limitations of the thermodynamic approach, the field of geochemical kinetics developed from the 1980s onwards (Lasaga 1998). Much progress was made studying mineral precipitation and dissolution kinetics as a function of solution composition (e.g. pH) and environmental conditions (e.g., temperature). These laboratory studies were done under well constrained conditions and in the absence of living organisms. However, application of these experimentally determined kinetic parameters to natural systems revealed that chemical kinetics often could not explain the differences between predictions based on chemical thermodynamics and kinetics, and observations in natural systems. These unfortunate discrepancies were attributed to the black box 'biology' or 'bugs'.

Before the molecular biology revolution, microbial ecology was severely method limited. Samples from the field were investigated using microscopy and total counts of bacteria were reported. Microbiologists were isolating a biased subset of microbes from their environment and studying their metabolic capabilities in the laboratory. To investigate whether these microbial processes occur in nature, microbial ecologists developed isotope and micro-sensor techniques to quantify rates of metabolism in natural environments (e.g., oxygen production or consumption, carbon fixation, sulfate reduction). These microbial transformation rates were of interest to geochemists because they represented the actual reaction rates, rather than the ones predicted from geochemical kinetics. Microbial ecologists and geochemists started to collaborate systematically and a new discipline emerged in which cross-fertilization of concepts, approaches and methods stimulated not only research questions at the interface but also in the respective disciplines. Stable isotope and organic geochemical biomarker techniques and detailed knowledge on mineral phases have enriched geomicrobiology, while knowledge on microbes and their capabilities and activities has advanced the understanding of elemental cycling. This integration of microbial ecology and geochemistry has evolved well regarding tools (e.g., the use of compound-specific isotope analysis and nanoSIMS in microbial ecology for identity-activity measurements), but less so in terms of concepts and theoretical development. Moreover, there is more to biology than microbiology. Animals and plants have a major impact on biogeochemical cycles, not only via their metabolic activities (primary production, nutrient uptake, respiration), but also via their direct impact on microbes (grazing, predation) and their indirect impact via the environment (ecosystem engineering: e.g., bioturbation, soil formation). This additional macrobiological component of biogeochemistry is increasingly being recognized (Middelburg 2018).

1.2 Focus on Carbon Processing in the Sea

This book focuses on biogeochemical processes relevant to carbon and aims to provide the reader (graduate students and researchers) with insight into the functioning of marine ecosystems. A carbon centric approach has been adopted, but other elements are included where relevant or needed; the biogeochemical cycles of nitrogen, phosphorus, iron and sulfur are not discussed in detail. Furthermore, the organisms involved in carbon cycling are not discussed in detail for two reasons. First, this book focuses on concepts and the exact identity of the organisms involved or the systems (open ocean, coastal, lake) is then less relevant. Secondly, our knowledge of the link between organism identity and activity in natural environments is limited. For instance, primary production rates are often quantified and phytoplankton community composition is characterized as well, but their relationship is poorly known. The extent of particle mixing by animals in sediments can be quantified and the benthic community composition can be described, but the contribution of individual species to particle mixing cannot be estimated in a simple manner.

The following chapters will respectively deal with production (Chap. 2) and consumption (Chap. 3) of organic carbon in the water column, the processing of organic carbon at the seafloor (Chap. 4), the impact of biogeochemical processes on inorganic carbon dynamics (Chap. 5), and the composition of organic matter (Chap. 6). The carbon cycle is covered using concepts, approaches and theories from different subdisciplines within ecology (phycologists, microbial ecologists and benthic ecologists) and geochemistry (inorganic and organics) and crosses the divides between pelagic and benthic systems, and coastal and open ocean. The book aims to provide the reader with enhanced insight via the use of very simple, generic mathematical models, such as the one presented in Box 1.1. Because of our focus on concepts, in particular the biological processes involved, there will be little attention to biogeochemical budgets and the role of large-scale physical processes in the ocean (Sarmiento and Gruber 2006; Williams and Follows 2011). Accurate carbon budgets are essential for a first-order understanding of biogeochemical cycles, but it is important to understand the mechanisms involved before adequate projections can be made for the functioning of System Earth and its ecosystems in times of change. To set the stage for a detailed presentation of biogeochemical processes, we first introduce a simple organic carbon budget for the ocean.

1.3   A 101 Budget for Organic Carbon in the Ocean

Establishing carbon budgets in the ocean, in particular during the Anthropocene, is a far from trivial task, involving assimilation of synoptic remote sensing and sparse and scarce field observations with deep insight and numerical modelling of the transport and reaction processes in the ocean. The important processes and thus flows of carbon in the ocean are related to primary production, export of organic carbon from the surface layer to ocean interior, deposition of organic carbon at the seafloor and organic carbon burial in sediments. Accepting 25% uncertainty, these numbers are well constrained at 50 Pg C y^{-1} (1 Pg or 1 Gt is 10^{15} gr) for net primary production, 10 Pg C y^{-1} for export production, 2 Pg C y^{-1} for carbon deposition at the seafloor and 0.2 Pg C y^{-1} for organic carbon burial (Fig. 1.1). Although no detailed, closed complete carbon budgets will be presented, estimates for individual processes, including gross primary production, chemoautotrophy and coastal processes, are presented in the following chapters. However, the 50-10-2-0.2 rule for carbon produced, transferred to the ocean interior, deposited at

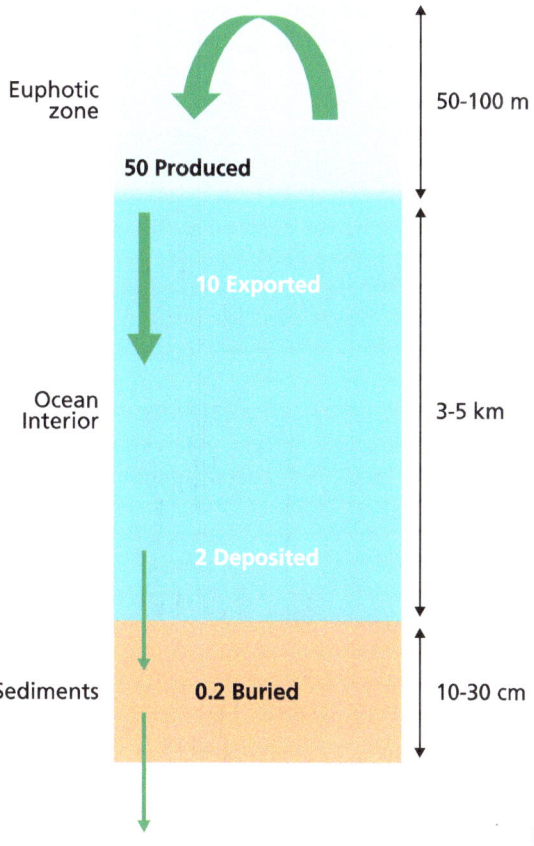

Fig. 1.1 Simplified budget of carbon flows in the ocean. Each year net phytoplankton production is about 50 Pg C (1 Pg = 1 Gt = 10^{15} g), 10 Pg is exported to the ocean interior, the other 40 Pg is respired in the euphotic zone. Organic carbon degradation continues while particles settle through the ocean interior and only 2 Pg eventually arrives at the seafloor, the other 8 Pg is respired in the dark ocean. In sediments, the time scale available for degradation increases order of magnitude with the result that 90% of the organic carbon delivered is degraded and only 0.2 Pg C yr^{-1} is eventually buried and transferred from the biosphere to the geosphere

Euphotic zone

50-100 m

50 Produced

10 Exported

Ocean Interior

3-5 km

2 Deposited

Sediments   **0.2 Buried**   10-30 cm

the seafloor and preserved in sediments, respectively, can easily be remembered and should be kept in mind when reading the details of carbon processing in the remaining of this book.

Box 1.1: A simple mathematical model for reaction and transport

In multiple chapters, we will make use of a very simple mathematical model in which the change in C (concentration, biomass) is due to the balance between diffusion (eddy Kz, molecular D), advection (sediment accretion particle/phytoplankton settling, w) and net effects of reactions (production and consumption). The basic equation is:

$$\frac{\partial C}{\partial t} = D\frac{\partial^2 C}{\partial x^2} - w\frac{\partial C}{\partial x} - kC + R_0$$

where $\frac{\partial C}{\partial t}$ is the change in concentration (mol m^{-3}) with time (t, s), $D\frac{\partial^2 C}{\partial x^2}$ is the spatial change in transport due to diffusion with diffusion coefficient D (m^2 s^{-1}), $w\frac{\partial C}{\partial x}$ is the spatial change in transport due to water flow or particle settling with velocity w (m s^{-1}), positive downwards, $-kC$ is the consumption of substance C via a first order reaction with reactivity constant k (s^{-1}) and R_0 is a zero-order production term (that is, the substance C has no impact on the magnitude of this rate).

This equation is based on spatially uniform mixing and settling rates and reactivity (i.e. D, w and k are constant). Moreover, we consider only steady-state conditions, i.e. there is no dependence on time. This simplifies the math: the partial differential equation $\left(\frac{\partial C}{\partial x}\right)$ becomes an ordinary differential equation $\left(\frac{dC}{dx}\right)$:

$$D\frac{d^2 C}{dx^2} - w\frac{dC}{dx} - kC + R_0 = 0$$

If we first consider the situation without zero-order production or consumption (i.e. $R_o = 0$), the general solution is:

$$C = Ae^{\alpha x} + Be^{\beta x}$$

$$\text{where } \alpha = \frac{w - \sqrt{w^2 + 4kD}}{2D} \text{ and } \beta = \frac{w + \sqrt{w^2 + 4kD}}{2D}$$

and A and B are integration constant depending on the boundary conditions. The number of integration constants sets the number of boundary conditions required. We will use models for the semi-infinite domain: i.e., if $x \to \infty$ then the gradient in C disappears ($\frac{dC}{dx} = 0$). Since all terms in β are positive, the second term becomes infinite and the integration constant B must thus be zero for this boundary condition.

For the upper boundary condition, we will explore two types: a fixed concentration and a fixed flux condition. If we know $C = C_0$ at depth $x = 0$, then A is C_0 and the solution is:

$$C = C_0 e^{\alpha x}.$$

Sometimes we know the external flux (F) of C, then we have to balance the flux at the interface at $x = 0$, e.g.:

$$F = -D\frac{dC}{dx}\Big|_{x=0} + wC|_{x=0}$$

Next, we take the derivative of the remaining first-term of the general solution $(Ae^{\alpha x})$, to arrive at:

$$F = -D\alpha Ae^{\alpha 0} + wAe^{\alpha 0}$$

Since $e^0 = 1$, $A = \dfrac{F}{-D\alpha + w}$ and the solution is:

$$C = \frac{F}{-D\alpha + w}e^{\alpha x}.$$

In some systems, transport is dominated by diffusion (e.g. molecular diffusion of oxygen in pore water, eddy diffusion of solutes and particles in water) and the advection term (w) can be ignored. The basic solutions given above remain but now $\alpha = -\sqrt{\dfrac{k}{D}}$ and the pre-exponential term for the constant flux upper boundary becomes $-\dfrac{F}{D\alpha}$. In other systems transport is dominated by the advection term (e.g. settling particles in the water column) and then $\alpha = -\dfrac{k}{w}$ and the flux upper boundary condition becomes $\dfrac{F}{w}$.

The above solutions are valid in the case that only first-order reaction occurs. The presence of zero-order reactions results in different solutions and these will be presented in the text where relevant. Similarly, the solutions presented are only valid if D, w and k are uniform with depth. In Chap. 3 we present an advection-first order degradation model in which we vary w and k with depth. Although user-friendly packages and accessible textbooks are available for numerical solving these and more complex equations (Boudreau 1997; Soetaert and Herman 2009), we restrict ourselves to analytical solutions because the relations among D, w and k in the various applications reveal important insights in the various process and governing factors, and the reader can implement the analytical solutions for further study.

References

Boudreau BP (1997) Diagenetic models and their implementation. In: Modelling transport and reactions in aquatic sediments. Springer, p 414

Lasaga AC (1998) Kinetic theory in the Earth Sciences. Princeton University Press, p 811

Lotka AJ (1925) Principles of physical biology. Wiliams & Wilkins, Baltimore, p 460

Lovelock JE (1972) Gaia as seen through the atmosphere. Atmos Environ 6:579–580

Middelburg JJ (2018) Reviews and syntheses: to the bottom of carbon processing at the seafloor. Biogeosciences 5:413–427

Redfield AC (1958) The biological control of chemical factors in the environment. Am Sci 46:205–221

Sarmiento J, Gruber N (2006) Ocean biogeochemical dynamics. Princeton University Press, 526 pp

Soetaert K, Herman, PMJ (2009) A practical guide to ecological modelling. Springer, 372 pp

Williams RG, Follows MJ (2011) Ocean dynamics and the carbon cycle. Cambridge University Press, p 404

Primary Production: From Inorganic to Organic Carbon

Primary production involves the formation of organic matter from inorganic carbon and nutrients. This requires external energy to provide the four electrons needed to reduce the carbon valence from four plus in inorganic carbon to near zero valence in organic matter. This energy can come from light or the oxidation of reduced compounds, and we use the terms photoautotrophy and chemo(litho)autotrophy, respectively. Total terrestrial and oceanic net primary production are each ~ 50–55 Pg yr^{-1} (1 Pg = 1 Gt = 10^{15} g; Field et al. 1998). Within the ocean, carbon fixation by oceanic phytoplankton (~ 47 Pg yr^{-1}) dominates over that by coastal phytoplankton (~ 6.5 Pg yr^{-1}; Dunne et al. 2007), benthic algae (~ 0.32 Pg yr^{-1}; Gattuso et al. 2006), marine macrophytes (~ 1 Pg yr^{-1}; Smith 1981) and chemo(litho) autotrophs (~ 0.4 and ~ 0.37 Pg yr^{-1} in the water column and sediments, respectively; Middelburg 2011). Much of the chemolithoautrophy is based on energy from organic matter recycling. Since, photosynthesis by far dominates inorganic to organic carbon transfers, we will restrict this chapter to light driven primary production.

Gross primary production refers to total carbon fixation/oxygen production, while net production refers to growth of primary producers and is lessened by respiration of the primary producer. Net primary production is available for growth and metabolic costs of heterotrophs, and it is the process most relevant for biogeochemists and chemical oceanographers. For the time being, we present primary production as the formation of carbohydrates (CH_2O) and ignore any complexities related to the formation of proteins, membranes and other cellular components (Chap. 6), because these require additional elements (nutrients). The overall photosynthetic reaction is:

$$CO_2 + H_2O + \text{light} \rightarrow CH_2O + O_2$$

J. J. Middelburg, *Marine Carbon Biogeochemistry*, SpringerBriefs in Earth System Sciences, https://doi.org/10.1007/978-3-030-10822-9_2

It starts with the absorption of light energy by photosystem II (PSII):

$$2H_2O + light \overset{PSII}{\rightarrow} 4H^+ + 4e^- + O_2$$

This reaction yields energy to generate adenosine triphosphate (ATP). The oxygen produced originates from the water and can be considered a waste product of photosynthesis. The protons and electrons generated subsequently react with nicotinamide adenine dinucleotide phosphate (NADP$^+$) at photosystem I (PSI):

$$NADP^+ + H^+ + 2e^- \overset{PSI}{\rightarrow} NADPH.$$

The energies of NADPH and ATP are then used to fix and reduce CO_2 to form carbohydrate.

$$CO_2 + 4H^+ + 4e^- \overset{RuBisCO}{\rightarrow} CH_2O + H_2O$$

This reaction is normally mediated by the enzyme ribulose bis-phosphate carboxylase (RuBisCO).

Primary production is at the base of all life on earth; it is thus important to quantify it and to understand the governing factors. We will first present, at a very basic level, the primary producers. This will be followed by the introduction of the master equation of primary production, based on laboratory studies, and then a discussion of its application to natural systems.

2.1 Primary Producers

Primary producers in the ocean vary from μm-sized phytoplankton to m-sized mangrove trees. Phytoplankton refers to photoautotrophs in the water that are transported with the currents (although they may be slowly settling). Biological oceanographers usually divide plankton (all organisms in the water that go with the current) into size classes (Table 2.1). Most phytoplankton are in the pico, nano and microplankton range (0.2–200 μm). The prefixes pico and nano have little to do with their usual meaning in physics and chemistry. Their small size gives them a high-surface-area-to-volume ratio which is highly favourable for taking up nutrients from a dilute solution. Within these phytoplankton size classes there is high diversity in terms of species composition and ecological functioning. Both small cyanobacteria (Synechococcus and Procholoroccus) and very small eukaryotes (e.g., Chlorophytes) contribute to the picoplankton. Microflagellates from various phytoplankton groups (Chlorophytes, Cryptophytes, Diatoms, Haptophytes) dominate the nanoplankton and differ in many aspects (cell wall, nutrient stoichiometry,

Table 2.1 Plankton size classes in the ocean

Size lass	Name (example)
<0.2 μm	Femtoplankton (virus)
0.2–2 μm	Picoplankton (bacteria, very small eukaryotes)
2–20 μm	Nanoplankton (diatoms, dinoflagellates, protozoa)
20–200 μm	Microplankton (diatoms, dinoflagellates, protozoa)
0.2–20 mm	Mesoplankton (zooplankton)
2–20 cm	Macroplankton

pigments, number of flagellae, life history, presence/absence of frustule). While phytoplankton communities can be described in terms of species, size classes or molecular biology data based partitioning units, they can also be divided into different functional types (diatoms because of Si skeleton, coccoliths with $CaCO_3$ skeleton, N_2-fixers, etc.). Unfortunately, taxonomic, functional and size partitionings among phytoplankton groups are not necessarily consistent.

A substantial fraction of the ocean floor in the coastal domain receives enough light energy to sustain growth of photoautotrophs. This includes not only intertidal areas, but also the subtidal. Small-sized photoautotrophs (microphytobenthos, including diatoms and cyanobacteria) are again the dominant primary producers, but macroalgae, seagrass, saltmarsh plants and mangrove trees contribute as well. Seagrasses, saltmarsh macrophytes and mangrove trees have structural components and specialised organs (roots and rhizomes) to tap into nutrient resources within the sediments.

2.2 The Basics (For Individuals and Populations)

Carbon fixation by (and growth of) primary producers will be discussed based on the master equation of Soetaert and Herman (2009):

$$P = \mu \cdot B \cdot f_{lim}(\text{resources, conditions}) \tag{2.1}$$

This master equation simply states that production (P, mol/g per unit volume per unit time) is proportional to the biomass (B, mol/g per unit volume) of the primary producer, the actor, which has an intrinsic maximum growth rate of μ ($time^{-1}$) and is limited ($0 < f_{lim} < 1$) by either physical conditions (e.g., temperature, turbulence) or resources such as light, nutrients and dissolved inorganic carbon. This equation is simple and generic, and we will show below how it relates to phytoplankton global primary production estimates using remote sensing, to expressions used in numerical biogeochemical models and to exponential growth in the laboratory.

2.2.1 Maximum Growth Rate (μ)

Consider a primary producer in an experiment supplied with all the resources it needs and under ideal conditions, in other words the limitation function f_{lim} is equal to one and optimal growth occurs. Equation 2.1 then reduces to the change in B with time, or production P, is equal to $\mu \cdot B$:

$$P = \frac{dB}{dt} = \mu B \tag{2.2}$$

This is the well-known equation for exponential growth:

$$B = B_0 e^{\mu t}, \text{ or alternatively} : \mu = \frac{1}{t} ln \frac{B}{B_0} \tag{2.3}$$

where B is the biomass at times t and B_0 is the initial biomass. Plotting the logarithm of biomass development as function of time yields then a slope corresponding to μ. Sometimes data are reported as the number of cell divisions (or doublings) per day: $\mu_d = \frac{1}{t} log_2 \frac{B}{B_0}$.

Maximum growths for phytoplankton typically varies from 0.1 to 4 d^{-1}, implying doubling times $\left(\frac{ln2}{\mu}\right)$ of a fraction of a day to one week. Figure 2.1a shows a typical example of exponential growth for maximum growth rates of 0.1 to 2 d^{-1}. Exponential growth leads to rapid depletion of substrates and after some time, resources become limiting and phytoplankton enters into a stationary phase (Fig. 2.1b). Maximum growth size depends on phytoplankton group and size (Fig. 2.2; Box 2.1).

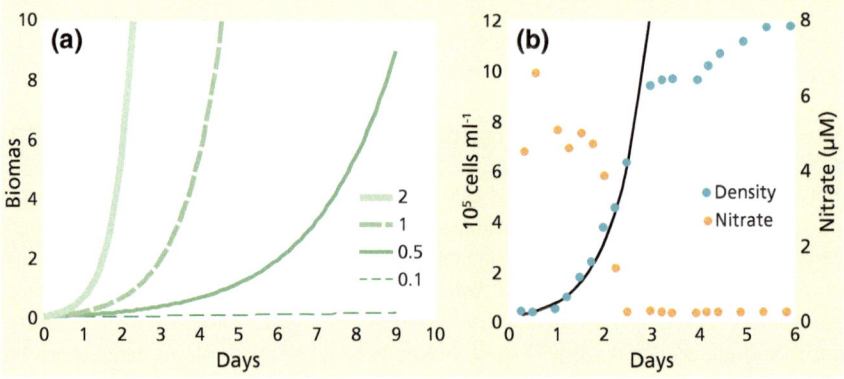

Fig. 2.1 a The increase in biomass during exponential growth with growth rates of 0.1, 0.5, 1 and 2 d^{-1}. **b** Cell growth of the diatom *Thalassiosira pseudonana* is exponential (growth rate of 1.4 d^{-1}) till nitrate is depleted and then stationary growth occurs (Data from Davidson et al. 1999)

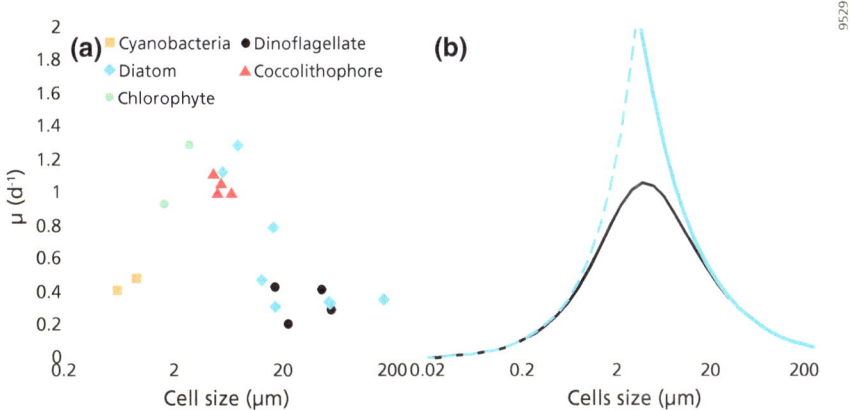

Fig. 2.2 a The relationship between growth rate and phytoplankton cell size and model prediction (grey curve) (Ward et al. 2017). **b** The relationship between growth rate and phytoplankton cell size (solid black line). Maximum nutrient uptake and requirement per cell scale positively with cell size (dashed blue line), while theoretical maximum growth rates scale negatively (solid blue line)

2.2.2 Temperature Effect on Primary Production

The temperature of a system provides a strong control on the functioning of organisms. Growth responses of populations to temperature are usually expressed by thermal tolerance curves, also known as reaction norms. Starting at low temperatures, growth initially increases linearly or exponentially up to a maximum T_{opt} and then typically declines relatively more rapidly: i.e. the response curve is often skewed to the left. In other words, phytoplankton growing near its optimum temperature is more sensitive to warming than to cooling (Fig. 2.3).

Although populations show distinct unimodal responses to temperature, mixed communities, and thus ecosystems, usually exhibit a smooth, monotonical increase best described by an exponential ($\mu = ae^{bT}$, Fig. 2.4). The thermal response can then be described by

$$\mu = ae^{bT}\left[1 - \left(\frac{T - T_{opt}}{width/2}\right)^2\right] \qquad (2.4)$$

where a and b are empirical parameters describing the maximum envelope for the mixed community and T_{opt} and *width* describe the maximum growth rate and temperature range of individual populations. Eppley's (1972) seminal work on temperature and phytoplankton growth in the sea reported values of 0.59 for a and 0.0633 for b. Note that this community response provides an upper limit for individual species and that high growth rates for individual species trade off with

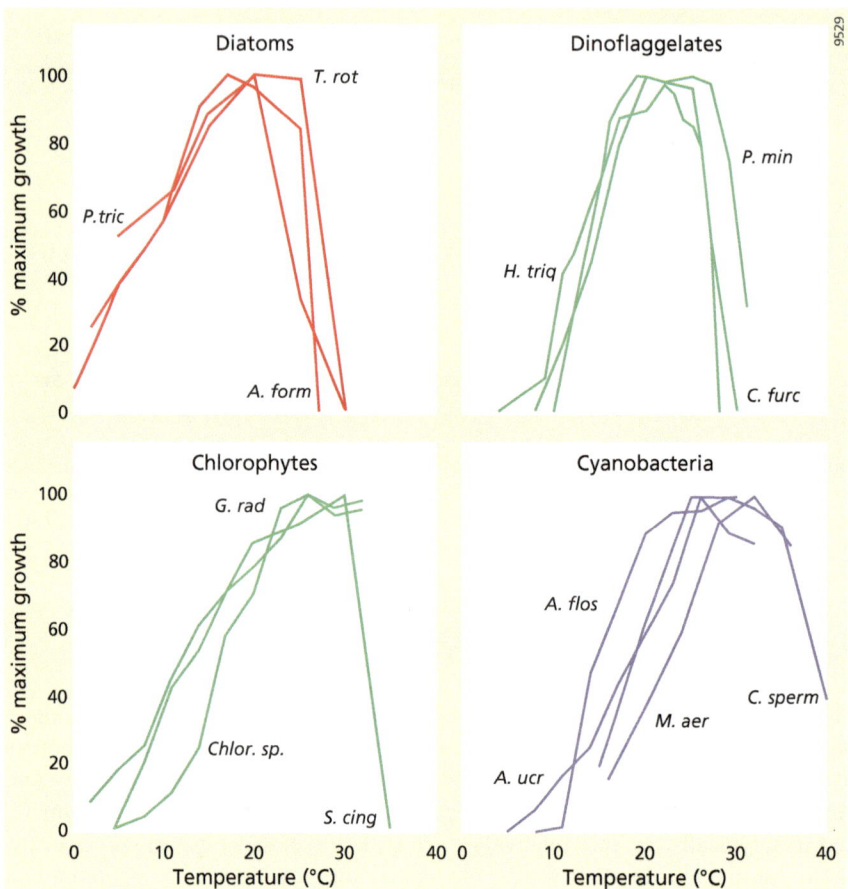

Fig. 2.3 Temperature response of the specific growth rates of three eukaryotic phytoplankton classes (Diatoms, Dinoflaggelates, Chlorophytes) and of Cyanobacteria from temperate freshwater and brackish environments (modified from Paerl et al. 2011). *A. form = Asterionella Formosa, T. rot = Thalassiosira rotula, P. tric = Phaeodactylum tricornutum, H. triq = Heterocapsa trique-tra, P. min = Prorocentrum minimum, C. furc = Ceratium furcoides, G. rad = Golenkinia radiate, Chlor. sp. = Chlorella sp., S. cing = Staurastrum cingulum, A. ucr = Anabaena ucrainica, M. aer = Microcystis aeruginosa, A. flos = Aphanizomenon flos-aquae, C. sperm = Cylindrospermopsis raciborskii*

growth rates at other temperatures, with the consequence that species replace each other (Fig. 2.4).

This exponential temperature response of natural communities is usually expressed in terms of Q_{10} values or Activation energies E_a, both rooted in chemical thermodynamics (van 't Hoff and Arrhenius equations). The temperature Q_{10} is normally defined as

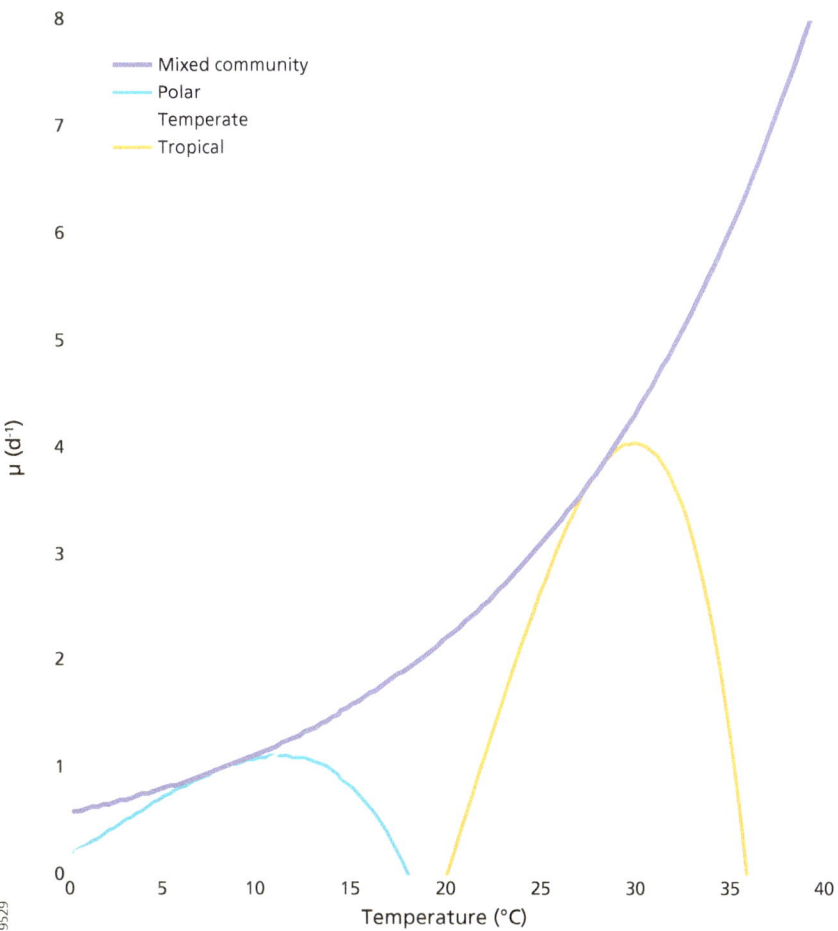

Fig. 2.4 Phytoplankton growth rate (d^{-1}) for a mixed community comprised of polar, temperate and tropical species. The mixed community response is based on Eppley (1972)

$$Q_{10} = \left(\frac{\mu_T}{\mu_{Ref}}\right) e^{\left[\frac{10}{\left(T - T_{Ref}\right)}\right]} \tag{2.5}$$

where μ_T and μ_{Ref} are the rate (e.g. growth) at temperature T and the reference temperature T_{Ref} (Celsius). Q_{10} can be simplified to

$$Q_{10} = \left(\frac{\mu_{Ref} + 10}{\mu_{Ref}}\right) \tag{2.6}$$

because it gives the rate increases for a 10 °C increase in T and is related to the parameter b of the exponential increase: $Q_{10} = e^{10b}$. Eppley's curve thus corresponds to a Q_{10} of 1.88. Typical Q_{10} values for biological processes are between 2 and 3.

The Arrhenius equation is very similar and reads

$$\mu = Ae^{\frac{-E_a}{RT}} \tag{2.7}$$

where A is a pre-exponential factor (time^{-1}), E_a is the activation energy (J mol^{-1}), R is the universal gas constant (8.314 J mol^{-1} K^{-1}) and T is the absolute temperature (K). Sometimes the universal gas constant R is replaced by the Boltzman constant k (8.617 10^5 eV K^{-1}) and then E_a is expressed in eV (energy per molecule) rather than J mol^{-1}. For the temperature range of seawater, E_a and Q_{10} values are related via

$$E_a = \frac{-R ln Q_{10}}{\left(\frac{1}{T} - \frac{1}{T_{Ref}}\right)} \text{ and } Q_{10} = e^{\left[\left(\frac{E_a}{R}\right)\frac{10}{\left(T \cdot T_{Ref}\right)}\right]}, \tag{2.8}$$

where T is again given in degrees Kelvin. Eppley's Q_{10} of 1.88 corresponds to activation energies of about 0.47 eV or 45 kJ mol^{-1} at 20 °C. One should realize that this is the optimal community temperature response, i.e. no other limiting factors. Apparent activation energies and Q_{10} values in the ocean are ~ 0.30 eV (29 kJ mol^{-1}) and ~ 1.5, respectively, close to that of Rubisco (Edwards et al. 2016).

2.2.3   Light

Photosynthesis is a light dependent reaction, and light intensity has a major impact on growth rates. The relationship between photosynthesis and irradiance is normally presented as a P versus E curve, where E refers to radiant energy (mol quanta m^2 s^{-1}). Multiple equations have been presented to represent the photosynthesis to light relation, which differ in the number of parameters and whether or not they include the photo-inhibition effect at high light intensities or respiration of the autotroph. Photorespiration, the breakdown of photo-labile, intermediate carbon fixation products, is important in full-light exposed organisms, such as terrestrial plants, microphytobenthos and phytoplankton in the surface layer.

Common simple limitation functions are the hyperbolic, Monod model:

$$f_{lim}(E) = \frac{E}{(E + K_E)} \tag{2.9}$$

where $f_{lim}(E)$ is the light limitation function ($0 < f_{lim}(E) < 1$), K_E is a light-saturation parameter (typically 50–150 µmol quanta m^{-2} s^{-1} for marine phytoplankton), and the Steele model (1962):

$$f_{lim}(E) = \frac{E}{E_{max}} e^{\left(1 - \frac{E}{E_{Max}}\right)} \tag{2.10}$$

where E_{max} is typically 50–300 µmol quanta m^{-2} s^{-1} for marine phytoplankton (Soetaert and Herman 2009). The Steele model represents both the initial increase and the subsequent decrease due to photo-inhibition with only one parameter (Fig. 2.5).

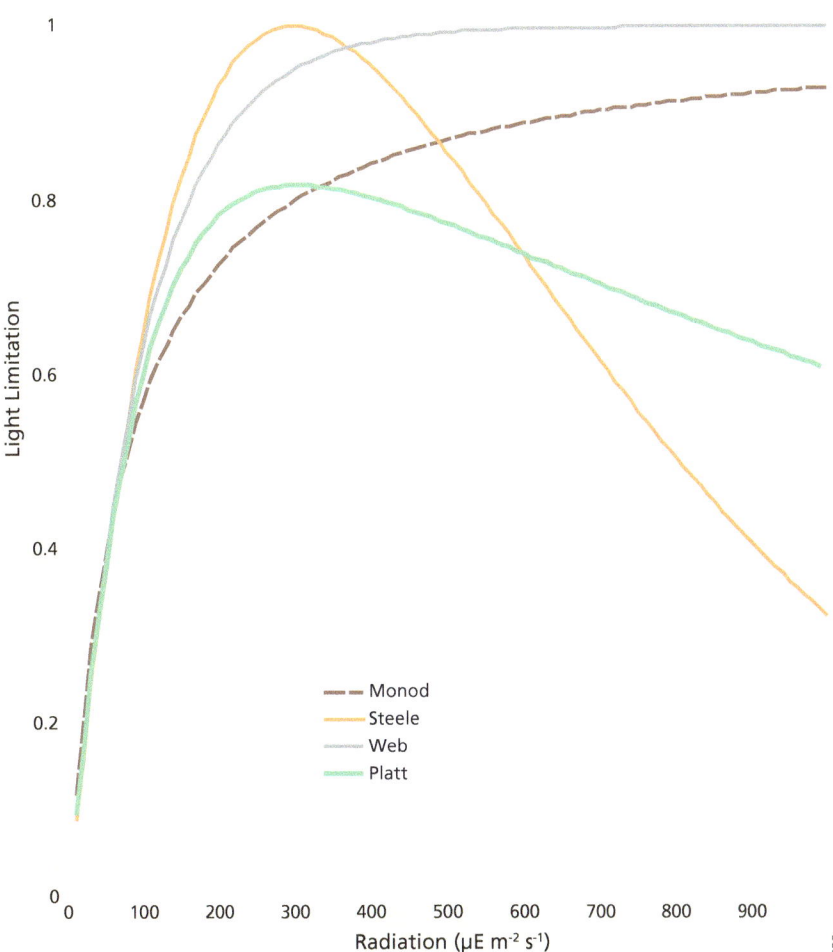

Fig. 2.5 Example of light inhibition functions in use. The Web and Monod models do not have light inhibition and only show different saturation behaviours, while the Steele and Platt models do incorporate the decrease in phytoplankton growth at high light levels, due to photo-inhibition

The Webb et al. (1974) model is based on an exponential:

$$f_{lim}(E) = 1 - \left[1 - e^{\left(\frac{\alpha E}{P_{max}}\right)}\right] \tag{2.11}$$

where P_{max} is the maximum rate at high light and α is the initial slope (increase in P with E at low light intensity). This equation ignores photo-inhibition. Alternatively, one can use the two-parameter Platt et al. (1980) equation:

$$f_{lim}(E) = 1 - \left[1 - e^{\left(\frac{\alpha E}{P_{max}}\right)}\right]\left[e^{\left(\frac{-\beta E}{P_{max}}\right)}\right] \tag{2.12}$$

where is β the intensity at the onset of photo-inhibition. Figure 2.5 illustrates the light limitation functions or PE curves presented above.

2.2.4 Nutrient Limitation

Growing phytoplankton needs a steady supply of resources to maintain growth. Nutrient uptake and growth kinetics are usually described using Monod or Droop kinetics. The former is the simpler model and normally used for steady-state conditions, while the Droop or internal quota model is preferred for transient conditions, e.g. in fluctuating environments. The equation for nutrient limitation following Monod kinetics is:

$$\mu = \mu_{max}\frac{S}{(S + K_{\mu})} \text{ or } f_{lim}(S) = \frac{S}{(S + K_{\mu})}, \tag{2.13}$$

where S is the substrate concentration of the medium water, $f_{lim}(S)$ is the nutrient limitation function, μ_{max} is the maximal growth rate, and K_{μ} is the half saturation constant for growth.

The Droop equation expresses growth rate as a function of the cellular quota (Q) of the limiting nutrient (Droop 1970):

$$\mu = \mu'_{max}\frac{Q - Q_{min}}{Q} \tag{2.14}$$

where Q_{min} is the minimum cellular quota for growth. Maximum growth rate on substrate (μ_{max}) and cellular quota (μ'_{max}) are related via $\mu_{max} = \mu'_{max}\frac{Q_{max} - Q_{min}}{Q_{max}}$ where Q_{max} is the maximum cellular quota if S increases.

2.3   From Theory and Axenic Mono-Cultures to Mixed Communities in the Field

Progress in theory, creativity in experimental design, and dedicated hard laboratory work has generated process-based understanding of phytoplankton growth in the laboratory. This body of knowledge has deepened our understanding and guided our modelling efforts and field observation strategies, but we need to make many assumptions before we can apply this mechanistic approach to the field.

Let us return to our master Eq. (2.1): $P = \mu \cdot B \cdot f_{lim}$ (resources, environmental conditions). Ignoring environmental conditions, such as temperature, and substituting the simplest expressions introduced above we arrive at:

$$P = \mu_{max} \cdot B \cdot \frac{E}{(E + K_E)} \cdot \frac{S}{(S + K_\mu)} \tag{2.15}$$

This equation for primary production contains 6 terms that need to be quantified for the case of a single limiting nutrient and a single phytoplankton species. The light availability (E) and nutrient concentration (S) display spatial and temporal gradients in nature, and the maximum growth rate μ_{max} and half-saturation dependences (K_E and K_μ) require experimental or laboratory studies.

2.3.1   Does Diversity Matter or Not?

One of the most critical restrictions on the use of mechanistic complex models is related to phytoplankton diversity. Hutchinson (1961) identified the paradox that phytoplankton is highly diverse, despite the limited range of resources they compete for, in direct contrast to the competitive exclusion principle (Hardin 1960). Seawater typically contains tens of different species of primary producers, many for which there are no maximum growth data and known limitation functions. Accordingly, it is not feasible to simply apply Eq. 2.15 to individual species in the field and sum their contributions to obtain the primary production. Besides these theoretical arguments against the single species approach, there are also empirical reasons. Primary production and its dependence on environmental conditions (nutrients, temperature, light) are normally quantified at the community level in the absence of techniques to quantify species-specific primary production in natural waters. This discrepancy between, on the one hand, mechanistic, single-species approaches in the laboratory and, on the other hand, quantification of community responses and activities is somewhat unfortunate (Box 2.2).

2.3.2 Chl the Biomass Proxy

The biomass of the primary producer (B) is the second term in our master equation and quantifying this term in natural systems is more difficult than one initially would anticipate. Particulate organic carbon (POC) concentrations (g C per unit volume) are a direct measure of phytoplankton biomass in laboratory settings with axenic cultures. However, in natural systems, the pool of particulate organic carbon comprises not only a mixture of phytoplankton species, each with its own maximum growth rate, temperature, light and nutrient dependence, but also a variable and sometimes dominating contribution of detritus (dead organic matter), bacteria and other heterotrophic organisms. It is for this reason that *chl*orophyll concentrations (*Chl*) are used as a proxy for living primary producer biomass. The rationale is that *Chl* is only produced by photosynthesizing organisms, degrades readily after death of the primary producers and can be measured relatively easily using a number of methods. Primary producer biomass (B) can then be calculated if one knows the C:*Chl* (or *Chl*:C) ratio of the phytoplankton. However, this ratio differs among species and depends on growth conditions, in particular light and nutrient availability (Cloern et al. 1995). *Chl*:C ratios vary from ~ 0.003 to ~ 0.055 (gC g*Chl*$^{-1}$; Cloern et al. 1995), complicating going from phytoplankton growth to primary production. The very reason that *Chl* is such a good proxy for photosynthesizing organisms is also the reason why it is not well suited to the task of partitioning itself among different phytoplankton species: it is in all primary producers harvesting light energy. Accessory and minor pigments such as zeaxanthine and fucoxanthine, do, however, have some potential to resolve differences among phytoplankton groups, but not at the species level.

2.3.3 Light Distribution

The distribution and intensity of photosynthetically active radiation in seawater is governed by the intensity at the sea surface (E_0) and scattering and absorption of light, with the result that light attenuates with depth. The decline of light intensity E with water depth z can be described by a simple differential equation, expressing that a constant fraction of radiation is lost:

$$\frac{dE}{dz} = -k_{PAR}E \tag{2.16}$$

where the proportionally constant k_{PAR} is known as the extinction coefficient (m^{-1}). Solving this equation using the radiation at the seawater-air interface (E_0) yields the well-known Lambert–Beer equation:

$$E = E_0 e^{-k_{PAR}z} \tag{2.17}$$

The extinction coefficient k_{PAR} includes the absorption of radiation by water (k_w), by the pigments from various primary producers (k_{Chl}), by coloured dissolved organic matter (k_{DOC}), and by suspended particulate material (k_{spm}). The light extinction coefficient of pure water ($k_w \approx 0.015\text{–}0.035$ m^{-1}) depends on the wave length of light, with longer wavelength (red) being adsorbed more strongly than shorter wavelengths (blue); this is the cause of the blue appearance of clear water. The other light extinction components have a different wavelength dependence: the attenuation coefficients of dissolved organic matter (k_{DOC}; "gelbstoffe") and detritus (k_{SPM}) increase with shorter wave length, while that of phytoplankton (k_{Chl}) varies depending on the species, i.e. the pigment composition of the primary producers (Kirk 1992; Falkowski and Raven 1997).

Oceanographers often divide ocean waters into two classes with respect to light absorption: case 1 waters in which phytoplankton (<0.2 mg Chl a m^{-3}) and its debris add only to k_w, and case 2 waters which have high pigment concentration and light attenuation because of (terrestrially derived) dissolved organic carbon and suspended particulate waters. The overall light attenuation (k_{PAR}) in case 1 waters can be approximated by (Morel 1988):

$$k_{PAR} = 0.121 \times Chl^{0.428} \tag{2.18}$$

where Chl is in mg Chl a m^{-3}.

Other useful empirical relations link light attenuation (k_{PAR}) to the Secchi depth (z_{Sec}, m), the depth at which a white disk disappears visually:

$$k_{PAR} = \frac{q}{z_{Sec}}, \tag{2.19}$$

where q varies from 1.7 in case 1 waters to 1.4 in case 2 waters (Gattuso et al. 2006) and

$$k_{PAR} = 0.4 + \frac{1.09}{z_{Sec}} \tag{2.20}$$

for turbid estuarine waters (Cole and Cloern 1987).

Light attenuation coefficients vary from 0.02 m^{-1} in oligotrophic waters, 0.5 m^{-1} in coastal waters, and to >2 m^{-1} in turbid waters Light attenuation by water and phytoplankton dominate in the open ocean and on the shelf. In other coastal waters, including estuaries, phytoplankton and suspended particles dominate light attenuation, while light attenuation is primarily due to suspended particles in more turbid systems (Heip et al. 1995).

The light attenuation governs the euphotic zone depth (z_{eu}, m), i.e., the depth where radiation is 1% of the incoming:

$$\ln 0.01 = -k_{PAR} z_{EU} \text{ or } z_{EU} = \frac{4.6}{k_{PAR}} \tag{2.21}$$

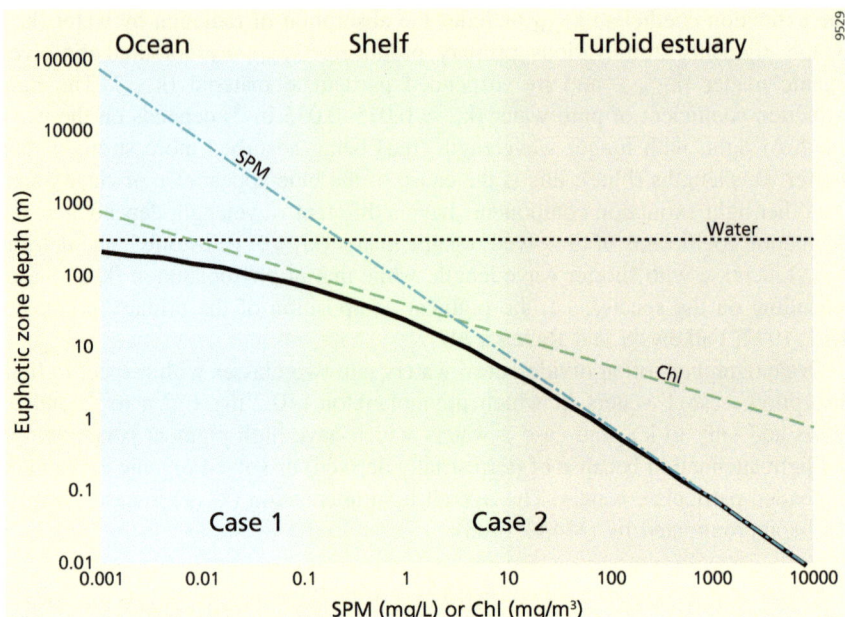

Fig. 2.6 Conceptual figure of euphotic zone depth (solid black line) as a function of suspended particulate matter (SPM, mg L^{-1}) and phytoplankton concentrations (*Chl*, mg m^{-3}). The euphotic zone is more >200 m in the clearest ocean water with very low phytoplankton and light attenuation by water itself dominates. In most of the ocean, phytoplankton dominates light attenuation and euphotic zone depth scales with phytoplankton concentration. In estuaries and other turbid systems, dissolved organic matter and in particular suspended particles attenuate light and the euphotic zone narrows to less than one meter. Coastal systems and eutrophic parts of the ocean are in between. Case 1 and 2 oceanic waters are indicated. Light attenuation due to phytoplankton was modelled following Morel (1988; Eq. 2.18), while that due to suspended particles followed Cloern (1987) and euphotic depth was calculated as 4.6/k$_{PAR}$

The euphotic zone is a key depth horizon in aquatic sciences because photosynthesis is largely limited to this zone. Moreover, the bottom of the euphotic zone is often used as reference for export of organic matter. Euphotic zone depths vary from about 200 m in the oligotrophic ocean, to tens of meters in shelf systems, to meters in coastal waters and a few decimetres in turbid and/or eutrophic estuaries (Fig. 2.6).

2.4 Factors Governing Primary Production

Having presented the factors governing phytoplankton production in laboratory studies and the limitations in applying that knowledge to natural systems, we have all the ingredients to explore the factors governing the (depth) distribution and rate of primary production in natural ecosystems.

2.4.1 Depth Distribution of Primary Production

Consider a system with a light profile following the Lambert–Beer equation (2.17) with $E_o = 10$ mol m^{-2} d^{-1} and $k_{PAR} = 0.1$ m^{-1} (corresponding to a euphotic zone of 46 m) and a nutrient pattern as shown in Fig. 2.7. Nutrients are low in the upper 25 m ($N = 0.1$ μmol m^{-3}) and then exponentially increase with a depth coefficient 0.1 m^{-1} to a maximum of 10 μmol m^{-3}.

If we further assume (1) that physical mixing homogenizes phytoplankton biomass (B = constant), (2) that there is only one limiting nutrients (N), and (3) that light and nutrient limitations can be described by Monod relations with parameter K_E and K_N. This allows combining μ_{max} and B into a depth independent maximal production P_m. The modelled P is then:

$$P = P_m \frac{E}{(E + K_E)} \cdot \frac{N}{(N + K_N)} \tag{2.22}$$

Taking K_N and K_E values of 1, i.e. 10% of E_o and maximum N at depth, and combining Eq. 2.22 with the light and nutrient profiles, we can then calculate the primary production as a function of depth (Fig. 2.7, green curve). Although these light and nutrient profiles and the model parameters K_N and K_E numbers have been chosen arbitrarily, they are reasonable and generate a representative depth profile for primary production with a subsurface maximum, as observed as a deep chlorophyll maximum (Fig. 2.8). In the upper 25 m, primary production is rather low because of nutrient limitation and declines slightly with depth because of light attenuation (Fig. 2.7). Primary production is optimal at depths between 25 and 40 m, i.e. where the nutricline and the lower part of the euphotic zone overlap. Primary production below 25 m is primarily light-limited, but accounts for about 75% of the depth-integrated primary production. Increasing surface-water nutrient concentrations or the phytoplankton affinity for nutrients (lowering K_N) would increase primary production in the top 25 m, but not so much at depth (Fig. 2.9a). Increasing the photosynthetic performance at low light levels (lowering K_E) would increase primary production at depth (Fig. 2.9b). Phytoplankton living in the surface ocean can thus optimize their performance by investing in nutrient acquisition, while those living in the subsurface would best optimize their light harvesting organs. This simple model explains why deep chlorophyll maxima occur in low-nutrient systems and why the depth distribution of primary production follows light in eutrophic systems (e.g. during early spring in Bermuda Atlantic station, Fig. 2.8).

2.4.2 Depth-Integrated Production

The overall control of light on depth-integrated production underlies satellite-derived algorithms for primary production and coastal predictive equations. For ecosystem and biogeochemical studies, the focus is on net primary production, i.e. carbon fixation minus phytoplankton respiration, expressed per m^2 and unit time

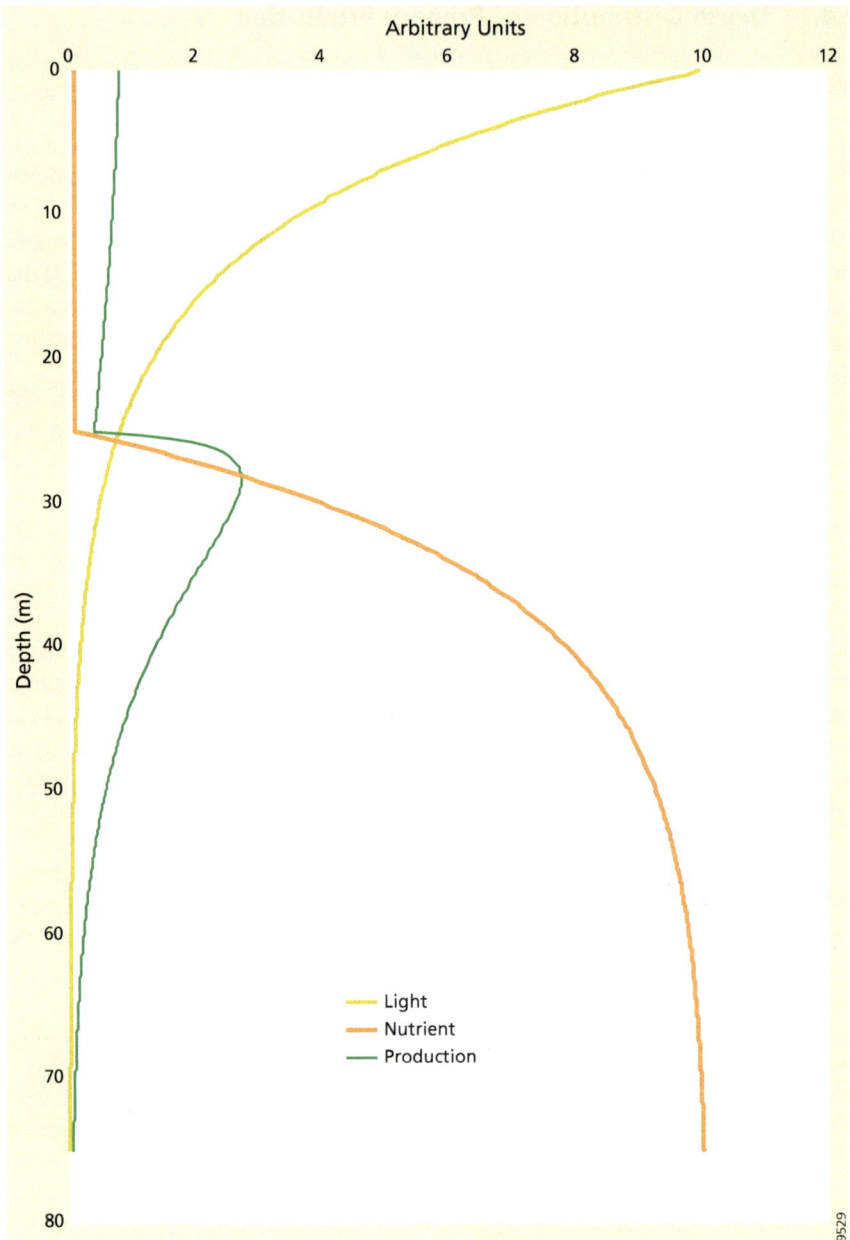

Fig. 2.7 Light and nutrient distribution versus depth and resulting primary production. The subsurface maximum often results in a deep chlorophyll maximum

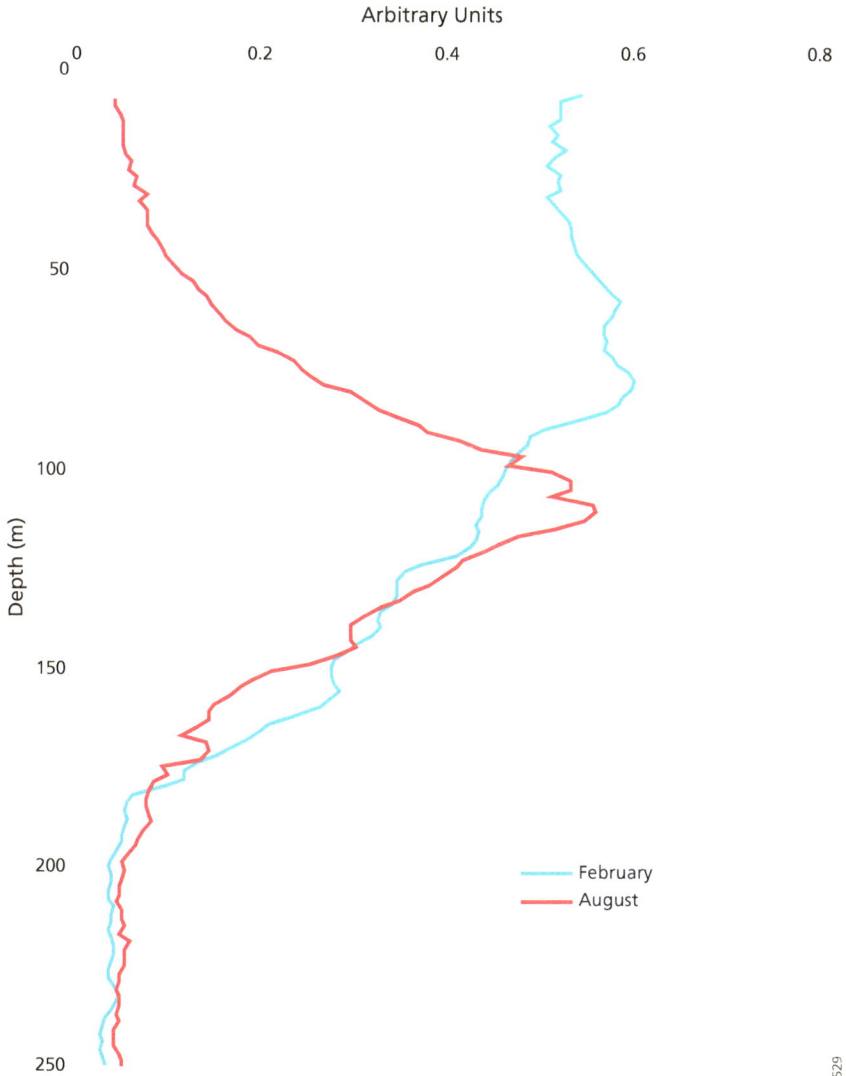

Fig. 2.8 Fluorescence data at station BATS (Bermuda Atlantic Time Series) in February and August 1996

(day, year). Behrenfeld and Falkowski (1997) showed that depth-integrated net primary production (P, g C m^{-2} yr^{-1}) can be estimated as:

$$P = P_{opt} \times Chl \times z_{eu} \times DL \times f_{lim}(E) \qquad (2.23)$$

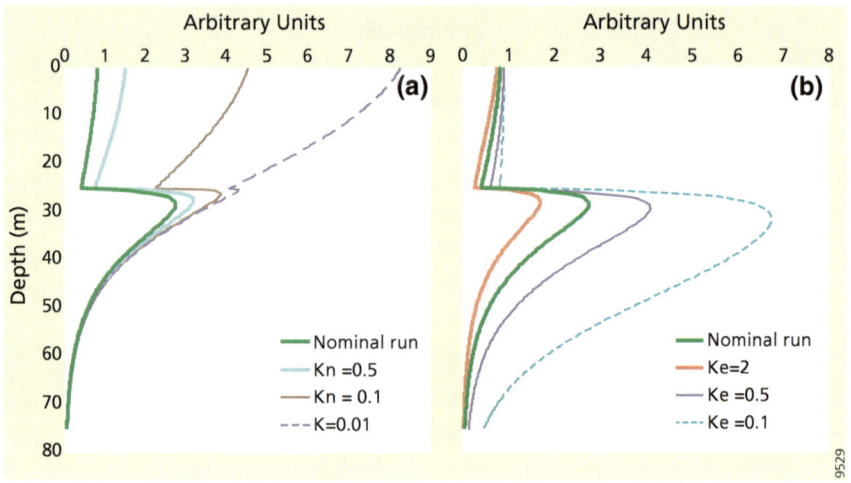

Fig. 2.9 a The impact of nutrient availability on the vertical distribution of primary production. Low half-saturation constants K_N imply high availability of nutrient for phytoplankton. **b** The impact of light harvesting efficiency on primary production. A high affinity (low K_E) for light causes higher primary production at depth. The nominal run is presented in Fig. 2.7

where P_{opt} is the maximum daily photosynthesis rate (mg C (mg *Chl*)$^{-1}$ h^{-1}), z_{eu} is the euphotic zone depth, DL is day length (h), and $f_{lim}(E)$ is a light limitation function. The similarity with our master Eq. (2.1) is evident, when nutrient limitation and environmental conditions are ignored. Integrating (Eq. 2.1) with depth to z_{EU}, and with time to sunset, we arrive at:

$$P = \int_0^{z_{EU}} \int_{sunrise}^{sunset} \mu_{max} \cdot B \cdot f_{lim}(E) \qquad (2.24)$$

which is identical to (2.23), with $P_{opt} = \mu_{max}$; *Chl* = B, $\int_0^{z_{EU}} = z_{eu}$, and $\int_{sunrise}^{sunset} = DL$.

Behrenfeld and Falkowski (1997) showed that 85% of the variance in global net primary production can be attributed to depth integrated biomass (*Chl* x z_{eu}) and the maximal photosynthesis parameter P_{opt}, with other factors, such as differences in light limitation functions, depth distributions of phytoplankton biomass and day length (DL), being less important. Consequently, the most rudimentary model would be (Falkowski 1981):

$$P = \psi \times Chl \times z_{EU} \times E_o \qquad (2.25)$$

stating that net primary production (P) scales linearly with depth integrated biomass (*Chl* x z_{EU}), incoming radiation (E_o) and an optimal photosynthetic parameter (ψ). Similar semi-empirical relations are often used in estuaries (Cole and Cloern 1987; Heip et al. 1995):

$$P = a + b(Chl \times z_{EU} \times E_0) \qquad (2.26)$$

where a and b are regression coefficients that are system specific.

2.4.3 Critical Depths

The overall governing role of light on primary production and phytoplankton dynamics also underlies the use of two critical depth horizons, often credited to Sverdrup (1953): the compensation depth (z_c) and critical depth (z_{cr}). These were introduced to understand and predict spring blooms in the ocean. At the compensation depth (z_c), phytoplankton photosynthesis is balanced by community respiration (Fig. 2.10), i.e. the depth of the radiation level at which photosynthesis (by phytoplankton) compensates their respiration (E_c). This compensation depth should not be confused with the physics governed mixed-layer depth (z_{mld}) and the critical depth (z_{cr}), where primary production integrated through the water column and over the day will equal the daily water column integrated community losses of carbon (Sverdrup 1953; Fig. 2.10). These depths are pivotal to the formation of

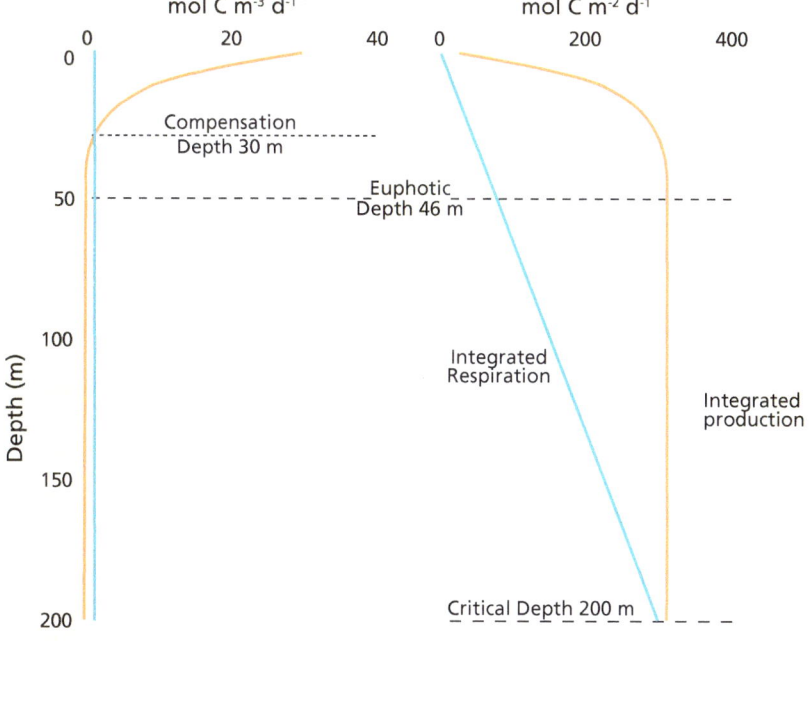

Fig. 2.10 Production and respiration (left) and depth integrated production and respiration (right) as a function of water depth. The critical depth is where depth integrated production and respiration balance (200 m), the compensation depth is where production and respiration of phytoplankton balance (30 m). The euphotic zone is governed only by light attenuation and is 46 m. P_0 is 30 mol m^{-2} d^{-1}; R_0 is $0.05*P_0$; $E_0 = 30$ mol m^{-2} d^{-1}; $k_{PAR} = 0.1$ m^{-1}

phytoplankton blooms in the oceans (Sverdrup 1953). If the mixed layer is deeper than the critical depth (z_{cr}), then phytoplankton will spend relatively too much time at low irradiances and carbon losses are not compensated by sufficient growth. Conversely, if the mixed layer is shallower than z_{cr}, phytoplankton communities can grow and blooms can develop. Assuming that carbon losses (R_o) are constant with depth, there is no nutrient limitation, and gross primary production is linearly related to radiation, which in turn depends exponentially on depth (Eq. 2.18), primary production is described by:

$P = P_0 e^{-k_{PAR} z}$, where P_0 is the surface productivity. One eventually arrives at following relations for Sverdrup's critical depth, z_{cr}:

$$\frac{\left(1 - e^{k_{PAR} z_{cr}}\right)}{k_{PAR} z_{cr}} = \frac{E_C}{E_0} = \frac{R_0}{P_0} \tag{2.27}$$

where E_c is the radiation level at the compensation depth and R_o is the depth-independent community respiration rate (Sverdrup 1953; Siegel et al. 2002). Clearly, light attenuation is a major factor, not only governing z_{eu}, but also z_c and z_{cr}. The critical depth (z_{cr}) is usually 4 to 7 times higher than the euphotic zone depth (z_{eu}). The compensation depth (z_c) is typically 50–75% of the euphotic zone depth (Siegel et al. 2002; Sarmiento and Gruber 2006; Fig. 2.10). For simplicity, the compensation depth is often taken equal to the euphotic zone depth; this should be discouraged, because it implies that community respiration represents only 1% of maximal production. The depth of the euphotic zone (z_{EU}) is an optical depth governed by the light attenuation and thus only indirectly impacted by phytoplankton via their effect on k_{PAR}, while the compensation depth depends on the community structure (algal physiology and heterotrophic community). The Sverdrup critical depth model is simple, instructive and predictive: it can explain bloom initiation when mixed layers shallow and link it to physical sensible and quantifiable parameters. However, it is sometimes difficult to apply because of inconsistencies and uncertainties in the parameterisation (phytoplankton vs. community respiration and other phytoplankton losses) and the validity of the assumptions (no nutrient limitation, well-mixed layer).

The critical depth horizon concept has been developed for deep waters, but a similar approach can be applied to shallow ecosystems. In shallow coastal systems, it is the relative importance of water depth and euphotic zone depth that governs (a) where production occurs and (b) whether phytoplankton biomass will increase or not. If water depth is less than the euphotic depth (z_{EU}) light reaches the seafloor and primary production by microbial photoautotrophs (microphytobenthos), as well as macroalgae and seagrasses, may occur. Gattuso et al. (2006) showed that this may happen over about 1/3 of the global coastal ocean. If water depth exceeds the euphotic zone by more than a factor 4–7 then phytoplankton losses in the dark cannot be compensated fully by photosynthesis and phytoplankton communities will lose biomass (Cloern 1987; Heip et al. 1995). Vice versa, if water depth <4–7 times z_{EU} phytoplankton growth is maintained. Consequently, shallowing of ecosystems (e.g. water flowing over a tidal flat or development of stratification)

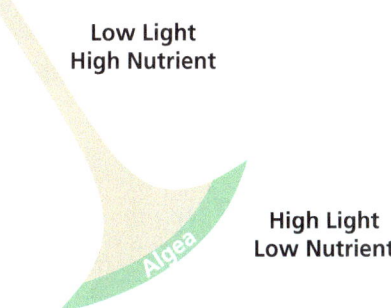

Fig. 2.11 Conceptual picture of phytoplankton bloom in estuarine plume

stimulates phytoplankton community growth, all other factors remaining equal, while deepening of water bodies will cause a decline. Moreover, in turbid systems where the light attenuation (k_{PAR}), and thus z_{EU} (Fig. 2.7), are governed by suspended particulate matter dynamics, phytoplankton communities may experience variable twilight conditions and have difficulty maintaining positive growth. Consequently, when turbid rivers and estuarine waters with high nutrients reach the sea, particles settle and light climate improves, phytoplankton blooms may develop and utilize the nutrients (Fig. 2.11).

Sverdrup's critical depth hypothesis is based on the assumption that phytoplankton biomass and phytoplankton losses are homogenously distributed in the mixed layer. However, the mixed layer with uniform temperature as used in Sverdrup's approach does not match with the layer of turbulent mixing in the ocean (Franks 2015). It is more realistic to represent phytoplankton biomass (B) as governed by the balance between production, respiration losses and transport by eddy diffusion and particle settling. Again, we assume gross primary production is linearly related to radiation (which declines exponentially); hence: $P = P_0 e^{-k_{PAR} z}$. Phytoplankton respiration loss is considered a first order process: $Loss = rB$ with a first-order rate constant (r). Under the assumption of steady-state we then arrive at (see Box 1.1):

$$K_z \frac{d^2 B}{dz^2} - w \frac{dB}{dz} - rB = P_0 e^{-k_{PAR} z} \qquad (2.28)$$

where K_z is the vertical eddy diffusion coefficient (m^2s^{-1}), w is the settling velocity ($m\ s^{-1}$; positive downwards), the other terms have been defined before. Considering a semi-infinite domain, i.e. $\frac{dB}{dx} = 0$ at large depth, and phytoplankton biomass B_0 at the water-air interface, we obtain the following solution:

$$B = \left(B_0 - \frac{P_0}{K_z k_{PAR}^2 + wk_{PAR} - r} \right) e^{\alpha z} + \frac{P_0}{K_z k_{PAR}^2 + wk_{PAR} - r} e^{-k_{PAR}z} \qquad (2.29)$$

with $\alpha = \frac{w - \sqrt{w^2 + 4rK_z}}{2K_z}$.

The second exponential term accounts for light-dependent production, while the first exponential comprises water-column mixing, phytoplankton settling, and phytoplankton losses. To simplify matters, we assume that phytoplankton biomass is zero at the air-water interface. The first and second term then balance if $\frac{w - \sqrt{w^2 + 4rK_z}}{2K_z} = -k_{PAR}$. After re-arrangement to isolate the eddy diffusion coefficient, we obtain

$$K_z = \frac{r - k_{PAR}w}{k_{PAR}^2} \qquad (2.30)$$

In other words, the vertical eddy diffusion coefficient K_z should be less than $\frac{r - k_{PAR}w}{k_{PAR}^2}$ for positive values of phytoplankton biomass (B).

Huisman et al. (1999) presented a more elaborate model on phytoplankton growth in a turbulent environment, including a feedback between phytoplankton biomass and k_{PAR}. Through scaling and numerical analysis of a model without phytoplankton sinking (w = 0), they derived a relationship between the maximum turbulent mixing coefficient K_z and k_{PAR}: $K_z = \frac{0.31}{k_{PAR}^2}$. If we also ignore phytoplankton advection (w = 0 in Eq. 2.30), $K_z < \frac{r}{k_{PAR}^2}$, fully consistent with Huisman et al. (1999). The critical turbulence level for phytoplankton growth is thus inversely related to the square of the attenuation of light. Moreover, the phytoplankton loss is the scaling factor. For turbid systems such as estuaries and other coastal systems with high light attenuation (k_{PAR}), turbulent mixing should be minimal to allow net growth, consistent with observations by Cloern (1991) that phytoplankton blooms develop during neap tide when turbulent mixing intensity is lowest. Conversely, in clear, oligotrophic waters, light attenuation is limited and phytoplankton blooms can occur at relatively high mixing rates. Sinking phytoplankton (w > 0) will lower the numerator of Eq. 2.30 and thus lower the critical turbulence levels, while buoyant phytoplankton (w < 0) will increase the maximal allowable turbulence, and thus the scope for phytoplankton growth.

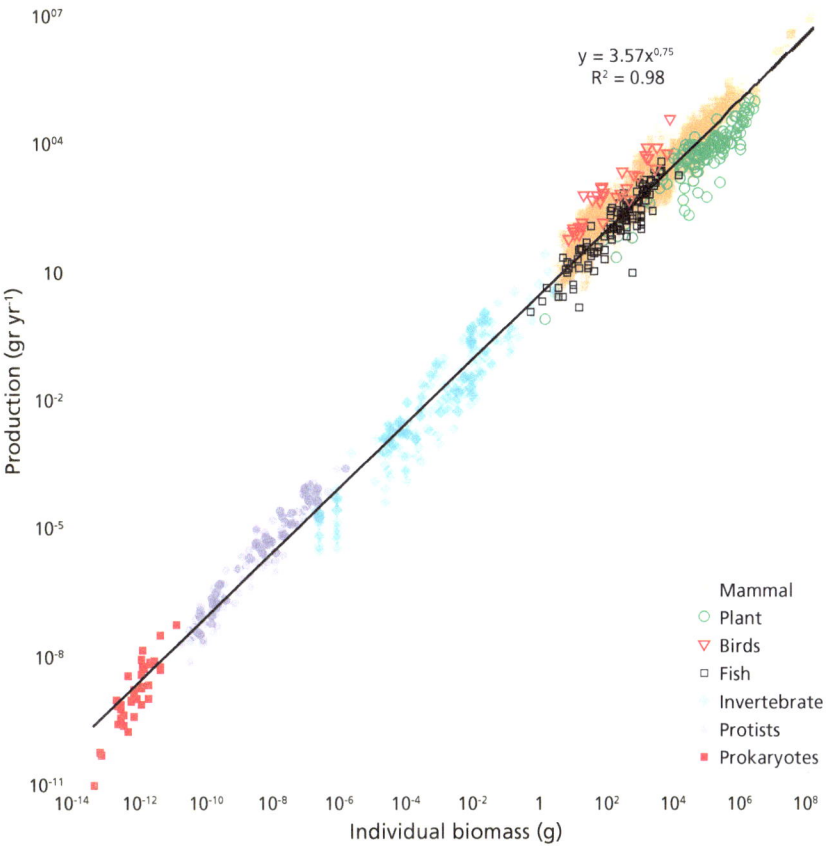

Fig. 2.12 Production as a function of individual biomass for a wide variety of organisms, including animals, plants and prokaryotes (based on data compiled by Hatton et al. (2015))

Box 2.1: Phytoplankton size based traits

The intrinsic maximum growth of phytoplankton varies with size (Fig. 2.2). Metabolic activity of organisms usually scales with size and when expressed in terms of mass or volume (V) follows a simple power law

$$\mu = aV^b, \tag{2.31}$$

where $b = -0.25$ according to the metabolic theory of ecology (Brown et al. 2004). Accordingly, the smaller the organism, the higher the intrinsic maximum growth rate. This power law relationship holds over orders of magnitude and across a wide range or organisms (autotroph and heterotroph, eukaryotes and prokaryotes; e.g., Fenchel 1973) and implies that smaller organisms have the highest intrinsic growth (Fig. 2.12).

However, some cell components are non-scaleable, such as the genome and membrane, and consequently this power-law appears to break down in the range of nanoplankton (2–20 μm). There is a trade-off between the size dependence of physiological traits (Ward et al. 2017). Burmaster's (1979) equation can be used to illustrate this:

$$\mu_{size} = \frac{\mu_{max} \cdot \theta_{size}}{\mu_{max} \cdot Q_{min} + \theta_{size}}, \tag{2.32}$$

where the maximum growth for a certain size (μ_{size}) depends on maximum nutrient uptake (θ_{size}), minimum cell quota (Q_{min}) and theoretical maximum growth rate (μ_{max}). Maximum nutrient uptake and requirement per cell scale positively with cell size (Fig. 2.2b, dashed blue line), while theoretical maximum growth rates scale negatively (Fig. 2.2b solid blue line). The result is an optimum in growth rate for phytoplankton in the nanoplankton range (Fig. 2.2b, black line). Very small picoplankton cells have a low intrinsic growth rate that will increase with size because more volume is then available for catalysing and synthesizing. The intrinsic growth rate of microplankton cells will decrease with increasing size, as with most organisms, for multiple reasons, including the increase in intracellular transport distances between cellular machineries (Marañón et al. 2013).

Box 2.2: Phytoplankton diversity, rate measurements and biogeochemical models

The high number of different species in each water sample poses a challenge to link the species-specific growth parameters obtained in the laboratory with measurements of phytoplankton growth in the field and modelling of phytoplankton primary production for natural, mixed communities. Gross primary production is normally quantified by the production of oxygen, using either ^{18}O-labelling or the differential evolution of oxygen in light and dark. The most common technique for measuring primary production is the ^{14}C labelling technique, but this method provides a result in between gross and net photosynthesis, depending on the duration of the incubation. Both approaches quantify primary production for the total community, rather than for specific species. Biological oceanographers have developed methods to quantify group-specific primary production, based on dilution approaches or the incorporation of isotopically labelled bicarbonate into biomarker or flow-cytometry separated groups of organisms (Laws 2013). These group-specific primary production measurements can be compared more directly to laboratory data.

Biogeochemical modellers have explored a number of strategies to incorporate differences among phytoplankton species into their ecosystem models; e.g. the plankton functional group approach and phytoplankton size or trait based approaches. The former approach is limited to a few plankton groups that are representative for certain biogeochemical fluxes (e.g. N_2-fixers, diatoms, small and large phytoplankton, coccoliths; Sarmiento and Gruber 2006). The size-based approach makes use of the systematic relationships between phytoplankton size and activity (e.g. Fig. 2.2), but some processes do not scale in a simple way with size. Trait- and genome-based approaches are the most recent, and they consider emergent phenomena (Follows et al. 2007). These approaches are instructive and needed to further our understanding and predictive capabilities in times of global change, but they are so far difficult to link with observations in the field.

References

Behrenfeld MJ, Falkowski PG (1997) A consumers' guide to phytoplankton primary production models. Limnol Oceanogr 42:1479–1491

Brown JH, Gillooly JF, Allen AP, Savage VM, West GB (2004) Toward a metabolic theory of ecology. Ecology 85:1771–1789

Burmaster DE (1979) The continuous culture of phytoplankton: mathematical equivalence among three steady-state models. Am Nat 113:123–134

Cloern JE (1987) Turbidity as a control on phytoplankton biomass and productivity in estuaries. Cont Shelf Res 7:1367–1381

Cloern JE (1991) Tidal stirring and phytoplankton bloom dynamics in an estuary. J Mar Res 49:203–221

Cloern JE, Grenz C, Vidergar LL (1995) An empirical model of the phytoplankton chlorophyll:-carbon ratio—the conversion factor between productivity and growth rate. Limnol Oceanogr 40:1313–1321

Cole BE, Cloern JE (1987) An empirical model for estimating phytoplankton productivity in estuaries. Mar Ecol Prog Ser 36:299–305

Davidson K, Wood G, John EH, Flynn KJ (1999) An investigation of non-steady-state algal growth. I. An experimental model ecosystem. J Plankton Res 21:811–837

Droop MR (1970) Vitamin B_{12} and marine ecology. V. Continuous culture as an approach to nutritional kinetics. Helgoländer Meeresun 20:629–636

Dunne JP, Sarmiento JL, Gnanadesikan A (2007) A synthesis of global particle export from the surface ocean and cycling through the ocean interior and on the seafloor. Glob Biogeochem Cycles, 21:GB4006. https://doi.org/10.1029/2006gb002907

Edwards KF, Thomas MK, Klausmeier CA, Litchman E (2016) Phytoplankton growth and the interaction of light and temperature: a synthesis at the species and community level. Limnol Oceanogr 61:1232–1244

Eppley RW (1972) Temperature and phytoplankton growth in the sea. Fish B.-NOAA 70:1063–1085

Falkowski PG (1981) Light-shade adaptation and assimilation numbers. J Plankton Res 3(2):203–216

Falkowski PG, Raven JA (1997) Aquatic photosynthesis. Blackwell

Fenchel T (1973) Intrinsic rate of natural increase: the relationship with body size. Oecologia 14:317–326

Field CB, Behrenfeld MJ, Randerson JT, Falkowski PG (1998) Primary production of the biosphere: integrating terrestrial and oceanic components. Science 281:237–240

Follows MJ, Dutkiewicz S, Grant S, Chisholm SW (2007) Emergent biogeography of microbial communities in a model ocean. Science 315:1843–1846

Franks PJS (2015) Has Sverdrup's critical depth hypothesis been tested? Mixed layers vs. turbulent layers. ICES J Mar Sci 72:1897–1907

Gattuso JP, Gentilli B, Duarte C, Kleypass J, Middelburg JJ, Antoine D (2006) Light availability in the coastal ocean: impact on the distribution of benthic photosynthetic organism and contribution to primary production. Biogeosciences 3:489–513

Hardin G (1960) The competitive exclusion principle. Science 131:1292–1297

Hatton IA, McCann KS, Fryxell JM, Davies TJ, Smerlak M, Sinclair ARE, Loreau M (2015) The predator-prey power law: biomass scaling across terrestrial and aquatic biomes. Science, 349: aac6284. https://doi.org/10.1126/science.aac6284

Heip CHR, Goosen NK, Herman PMJ, Kromkamp J, Middelburg JJ, Soetaert K (1995) Production and consumption of biological particles in temperate tidal estuaries. Oceanogr Mar Biol Ann Rev 33:1–150

Huisman J, van Oostveen P, Weissing FJ (1999) Critical depth and critical turbulence: two different mechanisms for the development of phytoplankton blooms. Limnol Oceanogr 44:1781–1787

Hutchinson GE (1961) The paradox of the plankton. Am Nat 95:137–145

Kirk JTO (1992) The nature and measurement of the light environment in the ocean. In: Falkowski PG, Woodhead AD, Vivirito K (eds) Primary productivity and biogeochemical cycles in the sea. Environmental Science Research, vol 43. Springer, Boston

Laws EA (2013) Evaluation of in situ phytoplankton growth rates: a synthesis of data from varied approaches. Ann Rev Marine Sci 5:247–268

Marañón E, Cermeño P, López-Sandoval DC, Rodríguez-Ramos T, Sobrino C, Huete-Ortega M, Blanco JM, Rodríguez J (2013) Unimodal size scaling of phytoplankton growth and the size dependence of nutrient uptake and use. Ecol Lett 16:371–379

Middelburg JJ (2011) Chemoautotrophy in the ocean Geoph. Res Lett 38:L24604. https://doi.org/10.1029/2011GL049725

Morel A (1988) Optical modeling of the upper ocean in relation to its biogenous matter content (case I waters). J Geophys Res 93:10749–10768

Paerl HW, Hall NS, Calandrino ES (2011) Controlling harmful cyanobacterial blooms in a world experiencing anthropogenic and climatic-induced change. Sci Total Environ 409:1739–1745

Platt TC, Gallegos L, Harrison WG (1980) Photo-inhibition of photosynthesis in natural assemblages of marine phytoplankton. J Mar Res 38(1980):687–701

Sarmiento J, Gruber N (2006) Ocean biogeochemical dynamics. Princeton University Press, p 526

Siegel DA, Doney SC, Yoder JA (2002) The North Atlantic spring phytoplankton bloom and Sverdrup's critical depth hypothesis. Science 296:730–733

Smith SV (1981) Marine macrophytes as a global carbon sink. Science 211:838–840

Steele JH (1962) Environmental control of photosynthesis in the sea. Limnol Oceanogr 7:137–150

Soetaert K, Herman PMJ (2009) A practical guide to ecological modelling. Springer, p 372

Sverdrup HU (1953) On conditions for the vernal blooming of phytoplankton. Journal du Conseil International pour l'Exploration de la Mer 18:287–295

Ward BA, Marañón E, Sauterey B, Rault J, Claessen D (2017) The size dependence of phytoplankton growth rates: a trade-off between nutrient uptake and metabolism. Am Nat 189:170–177

Webb WL, Newton M, Starr D (1974) Carbon dioxide exchange of Alnus rubra: a mathematical model. Oecologia 17:281–291

The Return from Organic to Inorganic Carbon

Almost all the organic matter produced is eventually consumed and respired to inorganic carbon because organic matter preservation via burial in accumulating sediments (~ 0.2–0.4 Pg y^{-1}) represents only a very small fraction of that produced. Global phytoplankton production is about ~ 50 Pg C y^{-1}, while phytoplankton biomass is ~ 1 Pg, implying a turnover of one week (0.02 y^{-1}). Marine macrophytes have a similar global biomass, but a production of only 1 Pg C y^{-1}; the turnover is thus ~ 1 y^{-1} (Smith 1981). These high turnover rates (compared with global terrestrial vegetation turnover of about one to two decades, Field et al. 1998) imply not only steady production, but also efficient consumption of organic matter. There are multiple organic matter loss pathways (respiration by autotrophs and heterotrophs, grazing, viral lysis, detrital route), but all eventually result in respiration and release of inorganic carbon.

Biogeochemists constructing carbon budgets normally lump together the various organic matter loss pathways and focus instead on the quantification of the organic carbon to inorganic carbon transformation. However, for a more detailed understanding, the elucidation of the link with other biogeochemical cycles and the identity of organisms involved, the various pathways have to be resolved. Before discussing the processes and mechanisms involved in these routes, it is instructive to distinguish between living organic matter that has the capability to reproduce (primary and secondary producers) and dead organic matter (i.e. detritus). While living organic matter is, by definition, fresh and thus labile, detrital organic matter pools represent a heterogeneous mixture of compounds from various sources which have been mixed together and which may have distinct compositions, degradation histories and reactivities (Box 3.1). Another distinction is between organic matter in the dissolved and particulate phases. Concentrations of DOC (order 10–100 mmol m^{-3}) are usually one order of magnitude higher than those of POC (order 1–10 mmol m^{-3}) in the euphotic zone of the ocean, while sediment POC (order 10^6 mmol m^{-3}) concentrations are normally three orders of magnitude higher than those of DOC in pore water (100–1000 mmol m^{-3}). Moreover, the DOC pool is detrital (with viruses), while the particulate organic carbon pool usually

© The Author(s) 2019

J. J. Middelburg, *Marine Carbon Biogeochemistry*, SpringerBriefs in Earth System Sciences, https://doi.org/10.1007/978-3-030-10822-9_3

represents a mixture of living organisms, their remains, and other detrital organic carbon inputs. As we discussed in Chap. 2, the POC concentrations in the water column can be a poor proxy for phytoplankton carbon. This distinction between dissolved and particulate pools is pivotal for understanding the fate of organic matter. Dissolved organic matter is transported as a solute with the water, while particulate organic matter is subject to gravity, which results in settling of organic particles. Consequently, the residence time of DOC in an aquatic system is much longer than that of POC. Consumption of dissolved organic matter primarily involves microbes and other small organisms, which use it for energy and nutrient acquisition, while larger organisms generally prefer particulate organic matter. Microbes use extracellular enzymes to solubilize particulate organic matter before they can utilize it.

In this chapter, we discuss carbon consumption in the context of the "biological pump". The latter depends on the fraction of primary produced organic matter that survives degradation in the euphotic zone and that is exported from surface water to the ocean interior, where it mineralized to inorganic carbon, with the result that carbon is transported against the gradient of dissolved inorganic carbon from the surface to the deep ocean. This transfer occurs through physical mixing and transport of dissolved and particulate organic carbon, vertical migrations of organisms (zooplankton, fish) and through gravitational settling of particulate organic carbon (Volk and Hoffert 1985; Sarmiento and Gruber 2006). We first discuss carbon consumption pathways in the euphotic zone, then the factors governing export of organic carbon from the euphotic zone and particle degradation in the ocean interior. Processes specifically related to organic carbon consumption in the coastal zone are presented in Box 3.2.

3.1 Carbon Consumption Pathway in the Euphotic Zone

For pelagic ecosystems, Legendre and Rassoulzadagan (1995) proposed a continuum of trophic pathways with the herbivorous food-chain and microbial loop as food-web end members (Fig. 3.1). The classical linear food-chain end-member involves grazing by zooplankton on larger phytoplankton and subsequent predation on zooplankton by either larger zooplankton or another predator. In such a linear food-chain a predator can either lead to high phytoplankton biomass (in a system with phytoplankton, herbivore and a predator) or reduced phytoplankton biomass (in a system with four levels). Changes in predator abundance can, thus, lead to trophic cascades (Pace et al. 1999). The microbial loop end-member involves not only phytoplankton, as basal resource, but also dissolved organic carbon (Azam et al. 1983). Dissolved organic carbon is used by heterotrophic bacteria for growth and respiration, and these heterotrophic bacteria are, in turn, consumed by micro-zooplankton (20–200 μm; ciliates, radiolarian, foraminifera) that are predated upon by larger zooplankton. Consequently, dissolved organic carbon is transformed, via a bacterial-microzooplankton loop, to zooplankton. These two end-member carbon processing pathways are connected at multiple levels. Small phytoplankton can be consumed directly by microzooplankton. Dissolved organic carbon is produced in multiple ways and by various organisms, both by primary producers and consumers of organic carbon (Fig. 3.1). DOC release by primary producers occurs passively by leakage and actively during unbalanced growth during nutrient limitation (Anderson and LeB Williams 1998; Van den Meersche et al. 2004). Another direct pathway from phytoplankton to dissolved organic pool involves viral lysis (Suttle 2005). Viruses are a major cause of phytoplankton mortality in the ocean, particularly in warmer, low-latitude waters. Sloppy feeding by herbivores and incomplete digestion of prey by consumers are other sources of dissolved organic carbon. Heterotrophic microbes use extracellular enzymes to solubilize particulate organic carbon and use this and other dissolved organic carbon resources for growth and maintenance. Part of the microbial heterotrophic production is used by microzooplankton; another part of the heterotrophic community is subject to intense viral lysis and this causes release of dissolved organic carbon again. The efficiency of the microbial loop depends on multiple factors but in particular on the relative importance of predation and viral lysis to the mortality of heterotrophic microbes.

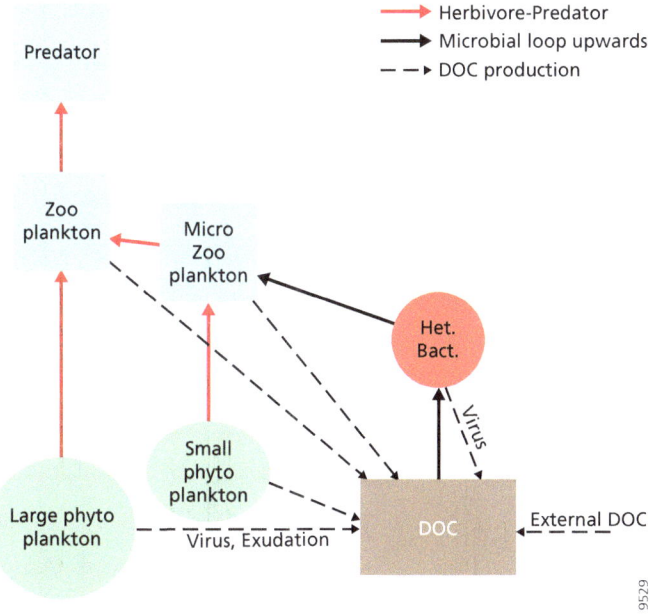

Fig. 3.1 Conceptual diagram of food web structure in euphotic zone. The linear food chain large phytoplankton-herbivore-predator (on the left with red arrow connections) has fewer levels than one with small phytoplankton at the base. The microbial loop refers to the flow from the dissolved organic carbon (DOC) via heterotrophic bacteria (Het. Bac.) and microzooplankton to predatory zooplankton (on the right with black solid arrows). Viruses play a major role in the mortality of phytoplankton and heterotrophic bacteria, and recycle organic carbon back to the DOC pool. Other sources of dissolved organic carbon (also dashed black arrows) includes exudation, sloppy feeding, etc. Particulate detritus pools and fluxes are not shown for simplicity

3.2   Factors Governing Export of Organic Matter

While net primary production (gross production minus respiration by the autotrophs) as presented in Chap. 2 is the property of interest for structure and functioning of food webs in the surface ocean, new and export production are most
relevant for the functioning of deep-sea and sedimentary ecosystems and the role of
the ocean in the global carbon cycle (Fig. 3.2). New production, also known as net
community or ecosystem production, refers to net primary production minus the
consumption of organic carbon by heterotrophs in the euphotic zone. At steady
state, this new production should be the same as what is exported, i.e. export
production. During periods of phytoplankton blooms new production may tem-
porarily exceed export out of the euphotic zone, while rates of export are tem-
porarily higher than new production during the senescence of phytoplankton
blooms.

The relations between export, new and net primary production are often
expressed in ratios. The e-ratio is defined as:

$$e-ratio = \frac{export\ production}{net\ primary\ production} \tag{3.1}$$

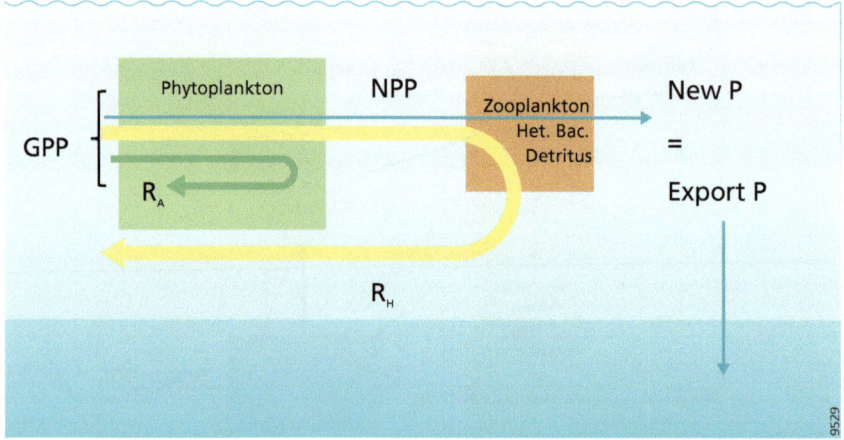

Fig. 3.2 The various types of productivity in the ocean. GPP (gross primary production) is the
total rate of carbon fixation/oxygen release by phytoplankton. Part of the GPP is used for
autotrophic respiration (R_A) by the phytoplankton community, the other part represents net
primary production (NPP). The organic matter produced is consumed by heterotrophs for growth
and respiration (R_H) and the remaining organic matter is new and available for export (NEP, new
production, export production). The net primary production (NPP) is most relevant for euphotic
zone food web functioning, while the export production is relevant for the biological pump and
fuelling deep-sea and benthic food webs. The export or e-ratio divides export/new production by
net primary production (NEP/NPP)

where export production is often quantified via the use of sediment traps (Suess 1980). Note that e-ratios based on sediment trap might be biased if dissolved organic matter is transported downwards by advection and/or eddy-diffusion and if zooplankton and other mobile vertical migrating organism contribute significantly to the total downward flux of organic carbon. An alternative method to quantify the relative importance of new to total production is the f-ratio, based on nitrogen use and recycling:

$$f\text{--}ratio = \frac{new\, production}{new\, production + recycled\, production} \tag{3.2}$$

where new production is assumed to equal nitrate uptake and recycled production is taken equal to ammonium uptake as determined from ^{15}N tracer experiments (Dugdale and Goering 1967). The rationale is that nitrogen availability limits primary production, that nitrate is supplied to the euphotic zone from deeper waters (thus new to euphotic zone) and that recycling of organic nitrogen within the euphotic zone generates ammonium. Export of organic nitrogen should eventually, on the global scale, be equivalent to nitrate supply from the deep to surface ocean. However, nitrification (ammonium oxidation to nitrite and nitrate) in the euphotic zone, atmospheric deposition, nitrogen fixation and other nitrogen cycling processes (bacterial uptake of dissolved inorganic nitrogen, dissolved organic nitrogen generation and use) complicate this simple depiction. At steady state e and f-ratios should be the same if the underlying assumptions are correct.

These export ratios vary from ~ 0.04 to ~ 0.72. Although multiple factors contribute to this range in export ratios (phytoplankton size, community structure, e.g. see Fig. 3.1), temperature and total primary production together account for 87% of the variance in export production and ratios (Laws et al. 2000, 2011; Dunne et al. 2005). The empirical model of Laws et al. (2011):

$$e\text{--}ratio = 0.04756\left(0.78 - \frac{0.43T}{30}\right)P^{0.307} \tag{3.3}$$

shows that e-ratios depend negatively on temperature (T, in Celsius) and scale with the power ~ 0.3 to net primary production (P). Export production consequently scales to net primary production with a power of ~ 1.3. Temperature is the single most important factor (Fig. 3.3), because organic carbon consumption by heterotrophs is more temperature sensitive than light and nutrient-limited primary production by phytoplankton (Laws et al. 2000). The impact of primary production on the export ratio can be explained by the increase in the size (and thus settling rate) of phytoplankton with increasing primary production (Dunne et al. 2005). The global average export ratio is ~ 0.2 with eutrophic, high-latitude systems having high e-ratios and oligotrophic, low-latitude systems having the lowest ratios.

3.3   Particulate Organic Carbon Fluxes in Ocean Interior

The efficiency of the biological pump depends not only on the rate of primary production (Chap. 2), the efficiency of carbon consumption within and export out of the photic zone (Fig. 3.3), but also on the depth at which the organic carbon is respired, because this determines the period during which carbon will be removed from the atmosphere (Yamanaka and Tajika 1996).

This pivotal role of carbon transfer to and within the ocean interior has stimulated observational programs to measure particle fluxes using sediment traps. Suess (1980), Pace et al. (1987) and Martin et al. (1987) were among the first to explore such sediment trap data for the global ocean, and they observed a steady decline of organic carbon fluxes with depth because of degradation during settling (Fig. 3.4a). The empirical open ocean composite from Martin et al. (1987) is most often used:

$$F_z = F_{z_0} \left(\frac{z}{z_0} \right)^{-b} \tag{3.4}$$

where z is water depth (m), z_0 is the reference level for export (e.g., bottom of the euphotic zone), F_{z/z_0} are the fluxes of organic carbon (mol m^{-2} d^{-1}) at depths z and z_0, and b is a fitted coefficient with a value of 0.858. Although we have made much

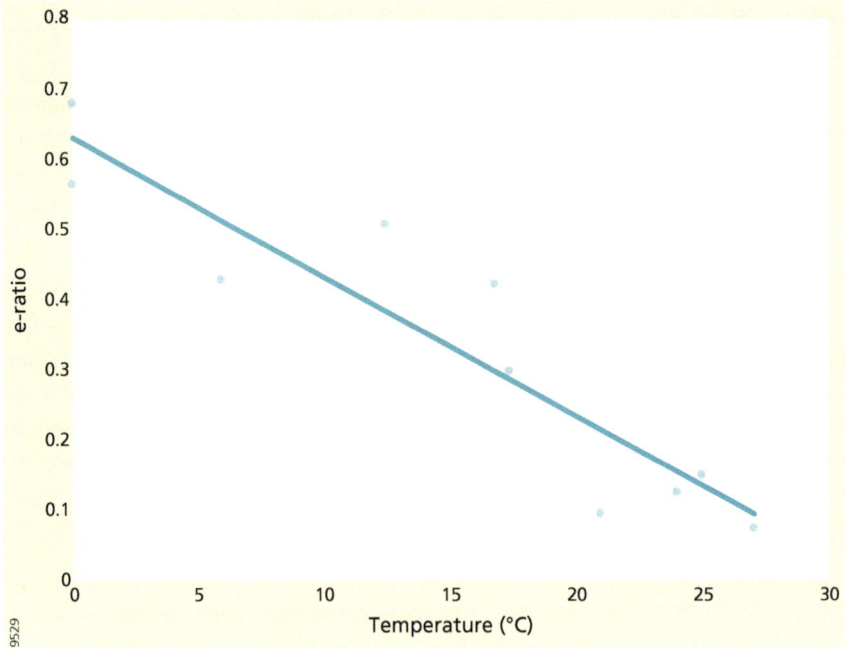

Fig. 3.3 Relation between e-ratio (export production over net primary production) as a function of water temperature (Data compiled by Laws et al. 2000)

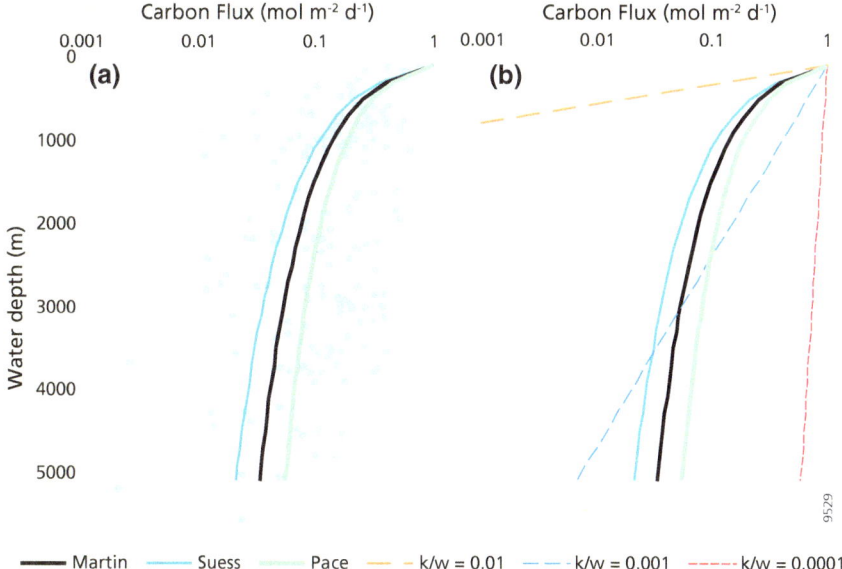

Fig. 3.4 a Fluxes of particulate organic carbon in the ocean interior based on the Martin et al. (1987), Suess (1980) and Pace et al. (1987) relations and data compiled by Lutz et al. (2002). Note that F_{z_0} has different meanings for the original equations but has been normalized to 1 for this plot. **b** Fluxes of particulate organic carbon in the ocean interior based on the Martin et al. (1987), Suess (1980) and Pace et al. (1987) relations and exponential curves with attenuation coefficient of 0.01, 0.001 and 0.0001 (m^{-1}). Martin et al.: $F_z = F_{z_0} \left(\frac{z}{z_0}\right)^{-0.858}$. Suess: $F_z = \frac{F_{z_0}}{(0.0238Z + 0.212)}$. Pace et al.: $F_z = F_{z_0} 3.523 z^{-0.734}$

progress in our understanding of the governing processes, the basic early findings have not been falsified. The open ocean composite from Martin et al. (1987) is used in many earth system and ocean biogeochemical models. The parameter b typically ranges from 0.6 to 1.3, with lower values in low oxygen settings and higher values in productive high-latitude systems (Berelson 2001). This variance in b values has been attributed to community structure, oxygen, temperature and mineral ballasting. Alternative more complex formulations have been derived.

To understand the factors governing particulate organic matter fluxes in the ocean interior, we will introduce a simple model for POC fluxes. At any water depth (z, directed downwards) below the export reference level (z_0), the particulate organic carbon concentration is governed by the balance of particulate organic matter supply by settling particles and degradation:

$$w\frac{dPOC}{dz} = -kPOC \tag{3.5}$$

where w is the settling velocity (m d^{-1}), POC is the concentration of particulate organic carbon (mol m^{-3}) and k is a first-order rate constant (d^{-1}). Assuming a constant first-order rate constant k (i.e. one-G model; Box 3.1), a constant settling rate (w), and a known export flux of organic carbon at the reference depth (F_{z_0}) the solution is an exponential:

$$POC_z = \frac{F_{z_0}}{w} e^{\frac{-k}{w}(z-z_0)} \qquad (3.6)$$

Moreover, $F_z = POC_z w$ because w is constant, and thus, the flux of particulate organic carbon is given by:

$$F_z = F_{z_0} e^{\frac{-k}{w}(z-z_0)} \qquad (3.7)$$

Accordingly, vertical organic carbon fluxes in the ocean are expected to decline exponentially with water depth and the attenuation is governed by the ratio between the organic matter degradation rate constant k and particle settling velocity w.

Figure 3.4b presents the Suess, Pace and Martin curves again but now together with three curves based on the exponential mechanistic model with k/w values varying from high attenuation (0.01) to very low attenuation (0.0001). While none of these can describe the observations, the former agrees with the upper part, while the latter agrees with the lower part. This was already noted by Martin et al. (1987) and implies that the assumption of constant k and/or w is likely not correct.

If our assumption of constant k is the main and only reason, then we would expect much improvement in model prediction with a multi-G or reactive continuum model (see Box 3.1). Figure 3.5 presents model predictions for both a multi-G and reactive continuum model based on model parameters for the phytoplankton decay experiments of Westrich and Berner (1984), presented by Boudreau and Ruddick (1991). While accounting for a decreasing reactivity of organic matter with depth improves the model performance in the upper part, there is still a large discrepancy at depth. However, better agreement can be obtained by fitting rather than imposing the organic matter degradation rate parameters based on literature values. The resulting gamma shape parameter (v > 0.6, depending on sinking velocity w chosen) is within the range of reported values (Arndt et al. 2013). Such a high v implies that settling organic matter is dominated by labile fractions, which seems reasonable.

Alternatively, the particle sinking velocity may be a function of water depth. Particle sinking velocity (w) is governed by Stokes' law:

$$w = \frac{2gr^2 \Delta\rho}{9\mu} \qquad (3.8)$$

where g is the gravitational acceleration (9.81 m^2 s^{-1}), r is the particle radius (m), $\Delta\rho$ is the density difference between seawater and the particle (kg m^{-3}) and μ is the dynamic viscosity of water (N s m^{-2}). In other words, particle settling is governed

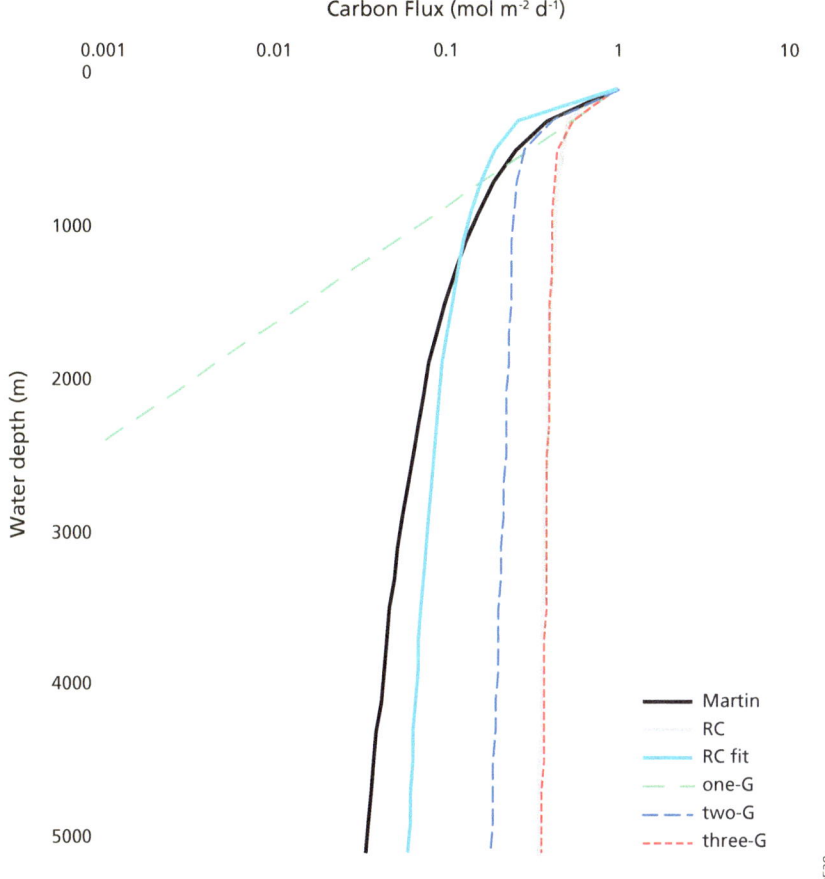

Fig. 3.5 Fluxes of particulate organic carbon in the ocean interior based on the Martin curve and model predictions using different parameterisation for organic matter degradation. The one-G has a fixed k of 0.03 (d^{-1}), the three G-model and reactive continuum (RC) parameters were taken from Boudreau and Ruddick (1991), the two G-model has similar k parameterisation as the three-G model but without a refractory fraction. The RC fit curve was based on fitting parameter v to the Martin curve (v = 0.59). Such a high v implies a dominance of fresh material

by the balance between gravitational acceleration of the particle ($g \times \Delta\rho$) and the drag acting upon it via surface area (r^2) and the friction of the fluid (μ). Particle size and density are the crucial factors (Sarmiento and Gruber 2006). The density difference $\Delta\rho$ is much larger for calcite and clay minerals with densities of ~ 2700 kg m^{-3} than for organic matter with a density of ~ 1060 kg m^{-3}, a little higher than that of water ~ 1027 kg m^{-3}. An organic particle with a diameter of 100 μm will settle ~ 12 m d^{-1}, while a pure calcite or clay mineral particles will sink ~ 600 m d^{-1}. Decreasing the size of the particle by a factor 10 lowers sinking velocities by a factor 100, because of the r^2 dependence, indicating that small reactive organic particles will never reach the seafloor.

Although Stokes' law is based on physical principles and generally valid, application to particles in the ocean is not straightforward. Particles are diverse and made up of aggregates of various plankton types, their detritus, faecal pellets and different mineral phases with distinct densities. Stokes' law as formulated above is for spherical particles and most marine particles, organism and their remains may deviate strongly from this shape, with the consequence that they are somewhat less dependent on size. Moreover, particles sizes in the ocean vary over orders of magnitude. Size spectra analysis show that most mass occurs in the smaller size classes and cannot directly be linked to that of primary producers size spectra. During settling, particles interact, disaggregate and coagulate, with the consequence that the number and size distribution of particles changes. Natural organic matter aggregates have high porosity, which decreases the density difference between water and particles and thus gravitational acceleration. Nevertheless, settling velocities of natural particles have shown to follow a power law with particle size (Clegg and Whitfield 1990; Sarmiento and Gruber 2006). In situ settling velocities have been quantified by high-resolution underwater video systems and inferred from the time-lag of pulse arrival between sediment traps at different water depth in the same area. The latter approach showed that particle settling velocities increased with depth (Berelson 2002).

To explore the impact of increasing settling velocities a linear and exponential increase of particle sinking rates with water depth was assumed (Fig. 3.6b) and Eq. 3.5 solved for these situations (Fig. 3.6a). A linear increasing rate of particle settling from 10 m d^{-1} at the bottom of the euphotic zone to about 185 m, combined with a first-order degradation rate constant (k = 0.03 d^{-1}; ~ 11 y^{-1}) agrees with the parameterisation of Suess (1980) and Martin et al. (1987). Evidently the initial assumption of constant first-order rate constants and sinking velocities might be questionable.

These agreements between the Martin's ocean composite model and the reactive continuum (with high v) and linear increasing settling rates are not unexpected: it is in the equations. The flux equation for the reactive continuum model with uniform w:

$$F_z = F_0 \left(\frac{aw}{aw + z} \right)^v \tag{3.9}$$

approaches the Martin et al. curve if aw $\ll$ z. Similarly, the flux equation for constant reactivity but with linear increasing velocity according to wz = c z (c in d^{-1}) is:

$$F_z = F_0 \left(\frac{z}{z_0} \right)^{\frac{-k}{c}} \tag{3.10}$$

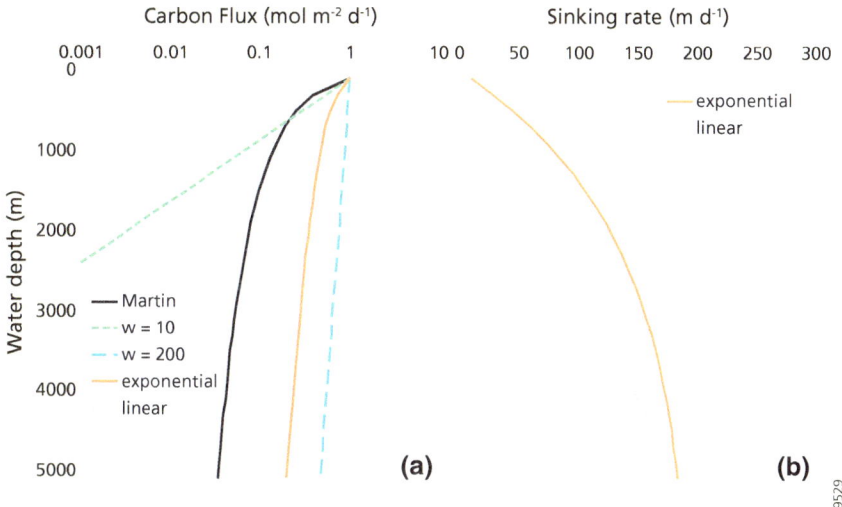

Fig. 3.6 a Fluxes of particulate organic carbon in the ocean interior based on the Martin curve and model predictions using different parameterisation for particle sinking rates. **b** Parameterisation for particle sinking rates in the ocean used to generate panel **a**. The solution to Eq. 3.5 with an exponentially increasing sink rate $w = v_0 + v_1 e^{-bz}$ is: $F_z = F_{z_0} e^{\frac{-k}{v_0}(z-z_0)} \left(\frac{v_0 + v_1 e^{-b(z-z_0)}}{v_0 + v_1} \right)^{\frac{-k}{v_0 b}}$ and with a linear increasing sink rate $w = c\, z$, it is: $F_z = F_{z_0} \left(\frac{Z}{Z_0} \right)^{\frac{-k}{c}}$

which returns Martin's power law if $\frac{-k}{c}$ is -0.858. With a k value of 0.03 d^{-1} (11 y^{-1}), settling velocities should then increase with a c value of 0.035 from 10 to about 185 m d^{-1} in the deep ocean, within the range of observed values (Berelson 2001).

Most of our knowledge of the nature, dynamics and magnitude of vertical organic particle fluxes in the ocean interior is based on sediment traps, which are one-side open vertical cylinders or conical containers that are fixed to a mooring or floating. These data may be biased because of hydrodynamics, degradation of material in the trap before recovery, and collection of swimming, foraging animals that are killed while feeding on the poisoned traps (Sarmiento and Gruber 2006).

An alternative approach to quantify the depth attenuation of particulate organic carbon fluxes in the ocean involves the use of sediments as the ultimate sediment trap. The idea is that eventually all organic material arriving at the seafloor is respired and can be measured by sediment oxygen uptake (the rationale for this approach will be explained in Chap. 4). Our mass-balance approach is shown in Fig. 3.7. Assuming that sediment respiration at a particular depth reflects the particulate organic carbon flux at that depth at any place in the ocean, we can calculate the total organic carbon flux as a function of depth by using the total area of the ocean at that depth. In other words, we assume a lateral homogenous ocean with no gradients in export between coastal and open ocean systems and between

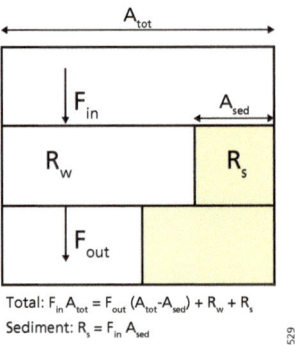

Fig. 3.7 Schematic picture for the carbon balance for a depth layer in the interior ocean. The incoming flux of settling carbon (F_{in}) over the total surface area (A_{tot}) of the ocean is balanced by the flux of carbon passing by (F_{out}) for the total surface area minus that of sediments (A_{sed}) and the carbon that is respired in the water column (R_w) and in the sediment (R_s). Moreover, sediment carbon respiration equals incoming flux (F_{in}) times global sediment area (A_{sed}) at that depth

high-latitude and low-latitude systems. Sediment oxygen consumption data of Andersson et al. (2004) were fitted with a power law and combined with ocean hypsometry (ETOP1) data. Moreover, by differencing carbon fluxes between two depth layers, we obtain total respiration in that depth zone. By substracting sediment respiration from total respiration we can derive the water-column respiration. A similar approach can be applied to the Suess and Martin relationships, as already shown by these authors. Water-column respiration rates estimated from sediment oxygen consumption are consistent with respiration rates calculated from the Martin et al. (1987) and Suess (1980) relations (Fig. 3.8), both in terms of profile shape, as well as magnitude. These are completely independent estimates: the sediment approach is based on sediment oxygen consumption, while the Martin and Suess approaches are based on primary production and sediment trap data. Depth integration of these global respiration versus depth profiles allows calculation of water-column respiration in the ocean interior (Table 3.1). Ocean interior water-column respiration estimates inferred from sediment oxygen consumption are intermediate between those derived from the Martin and Suess relationships (the latter is based on a global primary production of 50 Pg C y^{-1}), and consistent with literature estimates (Dunne et al. 2007). However, part of the settling carbon flux arrives at the ocean floor and this is not included in the Martin and Suess relationship. Total respiration in the ocean interior is, therefore, substantially higher (Table 3.1). While water-column respiration declines systematically with increasing water depth, global sediment respiration rates increase again below 1–2 km because of ocean hypsometry: large parts of the ocean have water depths between 3 and 6 km. As a consequence, at water depths more than 3 km sediments dominate organic carbon degradation, and thus oxygen consumption and carbon dioxide production (Fig. 3.8) This trend is robust because other sediment oxygen

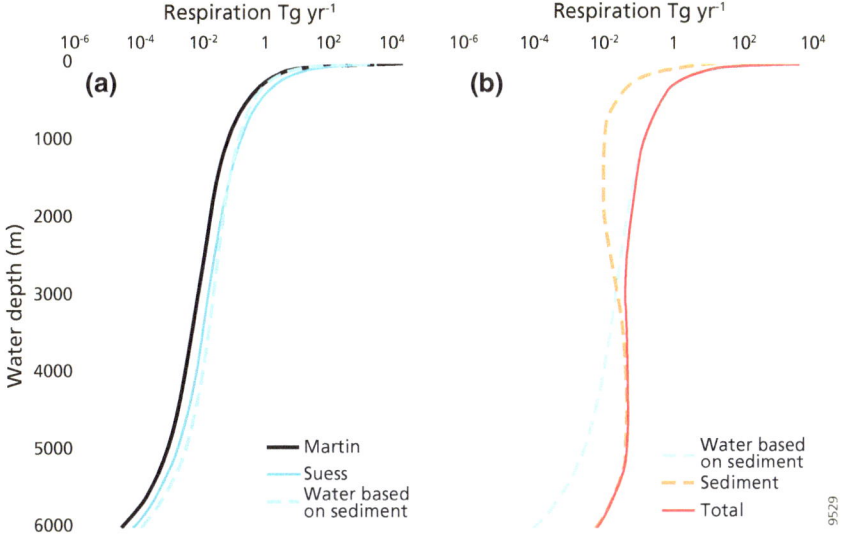

Fig. 3.8 a Water-column respiration in the ocean interior calculated from the Suess and Martin relationships and from sediment oxygen consumption data and the mass balance model of Fig. 3.7. 10 **b** Depth distribution of total respiration, water-column and sediment respiration based on a power fit to the Andersson et al. (2004) data. Sediment respiration dominates at water depths more than 3000 m because of hypsometry of the ocean. The Martin relations for respiration: $R_w = 13.1\left(\frac{Z}{100}\right)^{-1.858}$. The Suess flux relation $F_z = \frac{F_{z_0}}{(0.0238Z + 0.212)}$ was combined with a global net primary production of 50 Pg C y^{-1}. Sediment data are from Andersson et al. (2004) and were fit with a power law: $F_z = 84.88(z)^{-0.54}$ (mmol C m^{-2} d^{-1})

Table 3.1 Respiration in the ocean interior (Pg C y^{-1})

	Water column Martin[a]	Water column Suess[b]	Water column Rw[c]	Sediment Rs[c]	Total respiration[c]	Total Glud data[d]
100 m	5.85	17.40	8.79	1.77	10.56	14.07
200 m	3.05	8.64	5.53	1.59	7.12	8.26
1000 m	0.55	1.35	1.48	1.38	2.86	2.40
2000 m	0.19	0.45	0.60	1.26	1.86	1.36

[a]Water column respiration based on Martin et al. (1987)
[b]Water column respiration based on Suess (1980) and global primary production of 50 Pg C y^{-1}
[c]Total, sediment and water column respiration based on Andersson et al. (2004) data and approach outlined in Fig. 3.8
[d]Total respiration based on Glud (2008) sediment data and same as approach as for Andersson et al. (2004) data

consumption versus water depth curves generate similarly shaped curves and benthic contribution to deep-sea respiration (Table 3.1). Clearly, a more detailed look at sediment carbon processing is warranted, and this is provided in the Chap. 4.

Box 3.1: Organic matter reactivity

Detrital organic matter represents a very heterogeneous pool of thousands of organic compounds. Individual compounds may originate from various organisms, come from different environments, may have been produced recently or thousands of years ago, and have experienced a different history before eventually making up the compound pool of dissolved or particulate organic matter. Moreover, for particulate organic matter these compounds may be associated with different mineral phases or incorporated in various ways into minerals (Arndt et al. 2013). All these factors and intrinsic differences in reactivity among organic compounds result in large differences in organic matter reactivity (Middelburg 1989). Organic matter reactivity is usually expressed in terms of a first-order rate constant (k), which originates from the one-G model of marine organic matter degradation (Berner 1964):

$$\frac{dG_m}{dt} = -kG_m, \tag{3.11}$$

where G_m is the concentration of metabolizable organic matter, t is time and k is a first-order rate constant (time^{-1}) assumed to be constant in time. This equation implies that the concentration (G_m) and rate $\frac{dG_m}{dt}$ decrease exponentially with time. This model has been used successfully in various environments, from soils to sediments, from algal decomposition experiments to sewage degradation. However, there are two issues with its use (Middelburg 1989). One, it assumes that we know a priori the partitioning between refractory organic matter (G_r) and degradable organic matter (G_m) that add up to the total concentration (G). Two, organisms have been shown to preferentially utilize organic substrates, the more reactive being consumed first. This has a few consequences: the reactivity of the remaining organic matter decreases with reaction progress, i.e. time, and there are systematic changes in the composition of organic matter (Dauwe et al. 1999; Chap. 6).

To account for the observed decline in reactivity of organic matter with time, two alternative approaches have been presented: the multi-G and reactive continuum models. The multi-G model (Jørgensen 1978; Westrich and Berner 1984) divides the organic matter pool into a discrete number of fractions (usually two reactive and one refractory) with a different reactivity, each of which undergoes first-order decay. The relevant equations are:

$$G = \sum_i G_i \tag{3.12}$$

$$\frac{dG_i}{dt} = -k_i G_i \tag{3.13}$$

$$\frac{dG}{dt} = -\sum_i k_i G_i \tag{3.14}$$

where G_i is the concentration of organic carbon in each group i, k_i is the first-order reactivity of each group and i > 1. The one-G model (Eq. 3.11) is an example where i = 2, one labile class and one refractory group (with a k_i of zero). Selective removal according to the reactivity of each group accounts for the decrease in reactivity, amount and rate of organic matter degradation. However, successful application of this model requires knowledge on the number of labile groups, their contribution and their reactivity. These are unknown and cannot be measured.

The reactive continuum models do not subjectively partition organic matter into a number of pools but consider reactivity as an emergent property of the total organic matter pool that continuously declines as reaction progresses with time (Middelburg 1989). The two most commonly used models are the semi-empirical power model (Middelburg 1989):

$$\frac{dG}{dt} = -k(t)G, \tag{3.15}$$

where G is the total pool of organic matter and k(t) is time-dependent first-order rate constant, and the gamma-distribution reactive continuum model (Boudreau and Ruddick 1991):

$$\frac{dG}{dt} = -k_m G^{1+\frac{1}{v}} \tag{3.16}$$

where G is again the total pool of organic carbon, v is a parameter for the shape of the underlying Gamma distribution and $k_m = \frac{v}{aG_0^{\frac{1}{v}}}$ is the apparent rate constant for the mixture: a is a measure for average lifetime of the most reactive component and G_0 is the initial concentration of organic carbon. Moreover, reformulated in a closed form:

$$G = G_0\left(\frac{a}{a+t}\right)^v, \tag{3.17}$$

it describes the evolution of organic matter as a function of time (Boudreau and Ruddick 1991). The power and gamma-type reaction continuum model are under certain conditions equivalent and both are particular cases of the general q-theory of Bossata and Agren (1995). Although these continuum models are less used than the discrete models (Arndt et al. 2013), they have been applied for particulate organic matter in soils, sediments and suspended particles, phytoplankton degradation experiments and dissolved organic matter degradation in lakes. Apparent reactivity constants vary over 8 orders

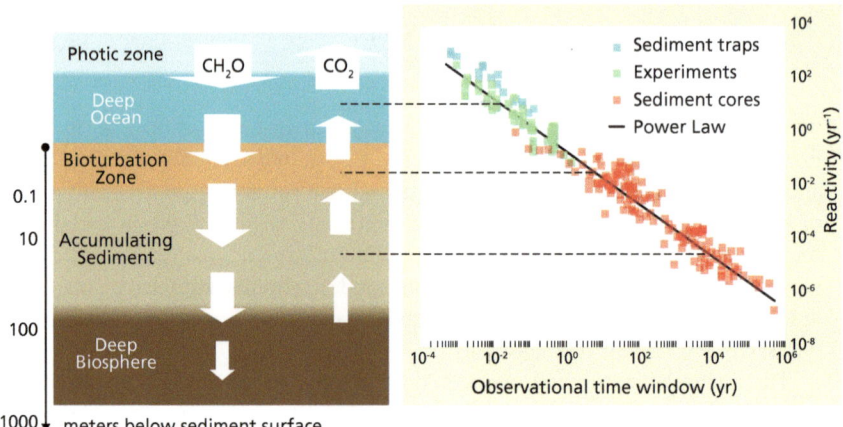

Fig. 3.9 The reactivity of particulate organic matter is an emergent property that decreases with the accumulated time period since it became detritus. Reactivity estimates are based on phytoplankton degradation experiments, settling phytoplankton and sediments in the accumulated zone. A power law ($k = 0.21t^{-0.985}$) relates k the first-order kinetic coefficient (yr^{-1}) with time (t) in yr (Middelburg 1989; Middelburg and Meysman 2007). Fresh phytoplankton initially has a high reactivity and this reactivity decreases during degradation because of preferential degradation, secondary production and mixing with more refractory components

of magnitude (Fig. 3.9), and we unfortunately lack analytical protocols to measure it directly or indirectly via a proxy. As a consequence, organic matter reactivity is a poorly constrained parameter in all biogeochemical models.

Box 3.2: Carbon consumption in the coastal ocean

In coastal zone, organic carbon is not only newly produced by phytoplankton, but also by microphytobenthos, macroalgae, seagrass, marshes and mangroves and imported from terrestrial ecosystems. Duarte and Cebrian (1996) presented the fate of autotrophic carbon based on an extensive cross-system survey (Fig. 3.10). Autotrophic respiration represents a loss of 26–35% for phytoplankton and microphytobenthos, while marine vegetations respire between 51 and 69% of their gross primary production. Herbivory losses are much higher for phytoplankton and microphytoplankton (26–37%), than for seagrass, marshes and mangroves (4–10%) with macroalgae in between ($\sim$16%). Detritus production and respiration were similar among these communities (15–27% of GPP) with the balance of organic material being exported to adjacent systems. These cross-system patterns in autotrophic respiration can be partly attributed to differences in biomass relative to production, because autotrophic respiration scales with biomass (Soetaert and

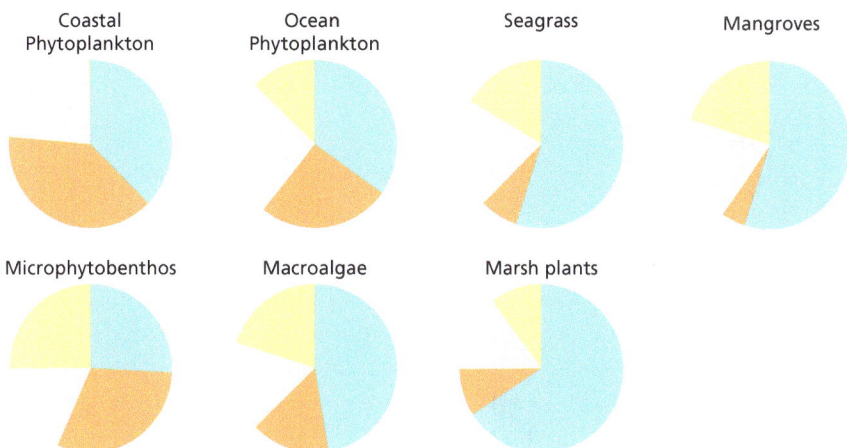

Fig. 3.10 Fate of primary produced materials for marine communities. Blue: autotrophic respiration, orange: herbivory; grey: degradation; yellow: export & accumulation (Duarte and Cebrian 1996)

Herman 2009). Differences in herbivory losses between phytoplankton and microphytobenthos on the one hand and higher plants on the other can be attributed to the general decline in growth rate and palatability and nutrient availability from microalgae to higher plants (Duarte and Cebrian 1996). The palatability of phytoplankton and microphytobenthos is generally much higher than that of terrestrial and littoral plants (mangroves, salt marshes) because the latter have lower nutrient and protein contents, higher proportions of structural components such as carbohydrates and lignins needed for rigidity and sometimes chemical compounds for defences. Grazers on seagrass and salt-marsh plants consequently have different feeding and digestion strategies than those on phytoplankton and microphytobenthos. While waterfowl, turtles and mammals such as the dugong are iconic grazers of seagrasses, many other organisms such as gastropods, isopods and sea urchins feed on epiphytes, the microbes growing on the leaves of seagrass. These epiphytes function more alike phytoplankton and microphytobenthos in terms of productivity, turn-over and their fate.

Herbivory represents a major loss pathway for phytoplankton and involves zooplankton as well as benthic suspension feeders. While benthic and pelagic grazers have many commonalities, there are also distinct differences: benthic suspension feeders are normally sessile, while planktonic float; benthic suspension feeders are usually larger and longer-lived with the consequence that they overwinter and already have a high biomass before the phytoplankton spring bloom. Moreover, benthic suspension feeders have only access to phytoplankton in the lower part of the water column and vertical mixing is an important factor governing their growth and their impact on phytoplankton dynamics (Herman et al. 1999).

Coastal systems exchange organisms and detrital material with adjacent ecosystems and receive detrital resources via rivers. This additional external organic matter is consumed and, as a consequence, the local balance between autotrophic and heterotrophic processes is disturbed, with net ecosystem heterotrophy as a result (Smith and Hollibaugh 1993). Heterotrophic ecosystems often show secondary production, similar to or higher than primary production, a dominance of microbial processes, a depletion in dissolved oxygen, and high carbon dioxide levels (Heip et al. 1995).

References

Anderson TR, LeB Williams PJ (1998) Modelling the seasonal cycle of dissolved organic carbon at station E1 in the English channel. Estuar Coast Shelf Sci 46:93–109

Andersson H, Wijsman JWM, Herman PMJ, Middelburg JJ, Soetaert K, Heip C (2004) Respiration patterns in the deep ocean. Geophys Res Lett 31. https://doi.org/10.1029/2003gl018756

Arndt S, Jørgensen BB, LaRowe D, Middelburg JJ, Pancost R, Regnier P (2013) Quantification of organic matter degradation in marine sediments: a synthesis and review. Earth Sci Rev 123:53–86

Azam F, Fenchel T, Field JG, Gray JS, Meyer-Reil LA, Thingstad F (1983) The ecological role of water-column microbes in the sea. Mar Ecol-Prog Ser 10:257–263

Berelson W (2001) POC fluxes into the ocean interior: a comparison of 4 US-JGOFS regional studies. Oceanography 14:59–67

Berelson WM (2002) Particle settling rates increase with depth in the ocean. Deep Sea Res II, 49:237–251

Berner RA (1964) An idealized model of dissolved sulfate distribution in recent sediments Geochimica et Cosmochimica Acta 28:1497–1503

Bossata E, Ågren GI (1995) The power and reactive continuum models as particular cases of the q-theory of organic matter dynamics. Geochimica et Cosmochimica Acta 59:3833–3835

Boudreau BP, Ruddick BR (1991) On a reactive continuum representation of organic matter diagenesis. Am J Sci 291:507–538

Clegg SL, Whitfield M (1990) A generalized model for the scavenging of trace metals in the open ocean. I.—Particle cycling. Deep-Sea Res 37:809–832

Dauwe B, Middelburg JJ, Herman PMJ, Heip CHR (1999) Linking diagenetic alteration of amino acids and bulk organic matter reactivity. Limnol Oceanogr 44:1809–1814

Duarte CM, Cebrian J (1996) The fate of marine autotrophic production. Limnol Oceanogr 41:1758–1766

Dugdale RC, Goering JJ (1967) Uptake of new and regenerated forms of nitrogen in primary productivity. Limnol Oceanogr 12:196–206

Dunne JP, Sarmiento JL, Gnanadesikan A (2007) A synthesis of global particle export from the surface ocean and cycling through the ocean interior and on the seafloor. Glob Biogeochem Cycles 21:GB4006. https://doi.org/10.1029/2006gb002907

Dunne JP, Armstrong RA, Gnanadesikan A, Sarmiento JL (2005) Empirical and mechanistic models for the particle export ratio. Global Biogeochem Cycles 19:GB4026. https://doi.org/10.1029/2004gb002390

Field CB, Behrenfeld MJ, Randerson JT, Falkowski PG (1998) Primary production of the biosphere: integrating terrestrial and oceanic components. Science 281:237–240

Glud RN (2008) Oxygen dynamics of marine sediments. Mar Biol Res 4:243–289

Heip CHR, Goosen NK, Herman PMJ, Kromkamp J, Middelburg JJ, Soetaert K (1995) Production and consumption of biological particles in temperate tidal estuaries. Oceanogr Mar Biol Ann Rev 33:1–150

Herman PMJ, Middelburg JJ, van de Koppel J, Heip CHR (1999) Ecology of estuarine macrobenthos. Adv Ecol Res 29:195–240

Jørgensen BB (1978) A comparison of methods for the quantification of bacterial sulfate reduction in coastal marine sediments. II. Calculation from mathematical models. Geomicrobiol J 1:29–47

Laws EA, D'Sa E, Puneeta N (2011) Simple equations to estimate ratios of new or export production to total production from satellite-derived estimates of sea surface temperature and primary production. Limnol Oceanogr Methods 9. https://doi.org/10.4319/lom.2011.9.593

Laws EA, Falkowski PG, Smith WO, Ducklow H, McCarthy JJ (2000) Temperature effects on export production in the open ocean. Glob Biogeochem Cycles 14:1231–1246

Legendre L, Rassoulzadegan F (1995) Plankton and nutrient dynamics in marine waters. Ophelia 41:153–172

Lutz M, Dunbar R, Caldeira K (2002) Regional variability in the vertical flux of particulate organic carbon in the ocean interior. Glob Biogeochem Cycles 16:1037. https://doi.org/10.1029/2000GB001383

Martin JH, Knauer GA, Karl DM, Broenkow WW (1987) VERTEX: carbon cycling in the Northeast Pacific. Deep Sea Res 34:267–285

Middelburg JJ (1989) A simple rate model for organic-matter decomposition in marine sediments. Geochim Cosmochim Acta 53:1577–1581

Middelburg JJ, Meysman FJR (2007) Burial at sea. Science 317:1294–1295

Pace ML, Cole JJ, Carpenter SR, Kitchell JF (1999) Trophic cascades revealed in diverse ecosystems. Trends Ecol Evol 14:483–488

Pace ML, Knauer GA, Karl DM, Martin JH (1987) Primary production, new production and vertical flux in the Eastern Pacific Ocean. Nature 325:803–804

Sarmiento J, Gruber N (2006) Ocean biogeochemical dynamics. Princeton University Press, Princeton, p 526

Smith SV (1981) Marine macrophytes as a global carbon sink. Science 211:838–840

Smith SV, Hollibaugh JT (1993) Coastal metabolism and the oceanic organic carbon balance. Rev Geophys 31:75–89

Soetaert K, Herman PMJ (2009) A practical guide to ecological modelling. Springer, Netherlands, p 372

Suess E (1980) Particulate organic carbon flux in the oceans-surface productivity and oxygen utilization. Nature 288:260–263

Suttle CA (2005) Viruses in the sea. Nature 437:356–361

Van den Meersche K, Middelburg JJ, Soetaert K, van Rijswijk P, Boschker HTS, Heip CHR (2004) Carbon–nitrogen coupling and algal–bacterial interactions during an experimental bloom: modeling a 13C tracer experiment. Limnol Oceanogr 49:862–878

Volk T, Hoffert MI (1985) Ocean carbon pumps: analysis of relative strengths and efficiencies in ocean-driven atmospheric CO_2 change. In: Sundquist ET, Broecker WS (eds) The Carbon cycle and atmospheric CO_2: natural variations Archean to present. AGU, Washington, pp 99–110

Westrich JT, Berner RA (1984) The role of sedimentary organic matter in bacterial sulfate reduction: the G model tested. Limnol Oceanogr 29:236–249

Yamanaka Y, Tajika E (1996) The role of the vertical fluxes of particulate organic matter and calcite in the oceanic carbon cycle: studies using an ocean biogeochemical general circulation model. Global Biogeochem Cycles 10:361–382

Carbon Processing at the Seafloor

4

Most of the global marine primary production (~ 55 Pg C y^{-1}) is consumed in the surface, sunlit ocean (~ 44 Pg C y^{-1}) and ocean interior (~ 8 Pg C y^{-1}) and only the remaining 2–3 Pg C y^{-1} reaches the sediment floor. Globally most of the organic carbon delivered to sediments ($\sim 90\%$) is degraded because organic carbon burial is low (~ 0.2–0.4 Pg C y^{-1}) and mainly occurs in rapidly accumulating coastal sediments.

In this chapter, we will first discuss organic matter delivery to the sediments, then the processes and organisms involved in organic matter degradation and close with a discussion of factors governing organic carbon burial.

4.1 Organic Matter Supply to Sediments

In Chap. 3 we have seen that organic carbon deposition on the seafloor depends primarily on water depth (Fig. 3.4), as a direct consequence of organic matter production in the euphotic zone and degradation of detritus during transit from the sunlit layer to the bottom of the ocean with more time for degradation in deep-water settings than in coastal settings. Consequently, deep-sea sediments receive little organic matter, while coastal sediments receive much. This water depth dependency of organic matter delivery rates is articulated by lateral differences in ocean primary production (production in shallow coastal systems is generally higher than that in the open ocean). Marine sediments are therefore often considered donor-controlled, i.e. the organic matter consuming benthic communities have no control over their food resources. While such a donor-controlled view applies to most deep-sea environments, there are other carbon delivery routes for sediments within the euphotic zone and for sediments inhabited by animals and these carbon transfers to sediments are (partly) mediated by the consuming communities, i.e. under consumer control (Fig. 4.1; Middelburg 2018).

© The Author(s) 2019
J. J. Middelburg, *Marine Carbon Biogeochemistry*, SpringerBriefs in Earth System Sciences, https://doi.org/10.1007/978-3-030-10822-9_4

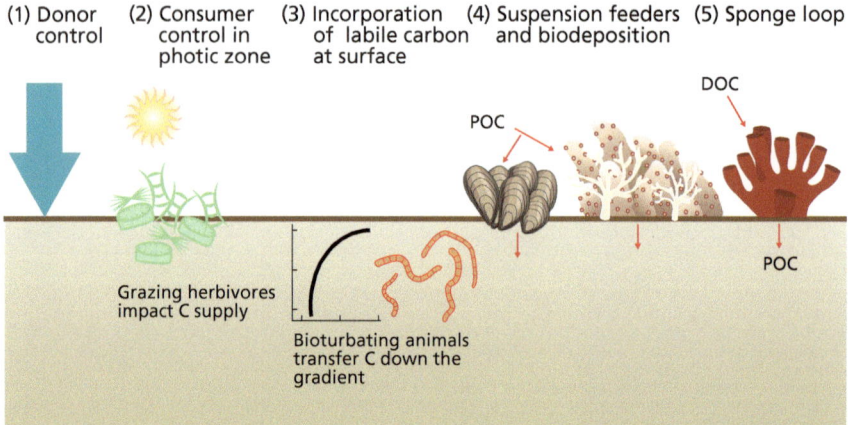

Fig. 4.1 Organic matter supply to sediments (Middelburg 2018). (1) Organic matter settling from the water column is deposited at seafloor (donor control; fixed flux upper boundary condition). (2) Sediments in the photic zone are inhabited by benthic microalgae that produce new organic matter in situ and grazing animals can impact the growth of these primary producers. (3) Bioturbating animals transfer labile carbon from the sediment surface layer to deeper layers in the sediments. (Vertical axis is depth; horizontal axis is concentration, see Fig. 4.2) (4) Suspension-feeding organisms enhance the transfer of suspended particulate matter from the water column to the sediments (biodeposition). (5) Sponge consume dissolved organic carbon and produce cellular debris that can be consumed by benthic organisms (i.e., the sponge loop)

Intertidal sediments and coastal sediments within the euphotic zone may support microbial photoautotrophs (microphytobenthos). Gattuso et al. (2006) showed that globally about 1/3 of sediments within the coastal ocean receive sufficient light for photosynthesis. This primary production at the seafloor represent a source of $\sim 0.32\,\mathrm{Pg\,C\,yr^{-1}}$ that is not only directly available for consumption, but it is also of high quality, because there has been no pre-depositional processing of organic matter. This benthic primary production input thus supports about 10–15% of global sediment respiration. Moreover, the grazing by micro, meio and macrofauna controls the biomass of the primary producers and in this way grazing animals control food supply for themselves, as well as for microbes living on the exudates and detritus of microphytobenthos (Middelburg et al. 2000; Middelburg 2018).

Macrophytes such as seagrasses in the subtidal euphotic zone and marsh plants and mangroves in the intertidal zone enhance carbon inputs to sediments in two ways. They locally produce new organic matter that is used for respiration, invested in leaves and stems, allocated to below-ground tissues for nutrient (and water) uptake, anchoring and storage. All this tissue can be consumed by microbes and animals, although the degradability of structural carbohydrate rich macrophyte material is often lower than that of protein-rich microphytobenthos and phytoplankton material (Duarte and Cebrian 1996; Fig. 3.10). Macrophyte communities stabilize the sediments and impact the local hydrodynamics with the consequence that suspended particles, including particulate organic matter, are trapped within their canopies. This trapped organic matter is often derived from adjacent ecosystems or phytoplankton from within the system and makes a substantial contribution to sediment organic carbon pools in salt marshes (Middelburg et al. 1997) and seagrass meadows (Gacia et al. 2002; Kennedy et al. 2010). Moreover, the trapped phytodetritus often is more readily degradable than locally produced macrophyte material (Barrón et al. 2006).

Suspension feedings worms, bivalves, corals and sponges consume particulate organic matter suspended in the water column. Part of the carbon is used for growth and maintenance, another part is excreted as faeces and pseudofaeces, which in turn are available for microbes and animals in the underlying and nearby sediments. Some organisms passively filter water, whereas others actively filter water to obtain their resources. Clearance rates (m^3 of water cleared from food per unit time) can be very high, and grazing by benthic suspension feeders often represents the largest loss term of phytoplankton in tidal systems (Heip et al. 1995). While some suspension feeders are spatially distributed rather homogenously, many others (e.g. oysters, mussels, coral reefs) occur in specific areas at high densities because of local favourable hydrodynamic conditions or positive feedbacks between community density, performance and local hydrodynamics (Herman et al. 1999). These suspension feeders locally enhance carbon input to underlying sediments relative to that delivered via sinking because they focus laterally available particulate organic carbon. Some of the highest carbon inputs to sediments have been reported for mussel and oyster beds. Moreover, benthic suspension feeders within the euphotic zone directly impact phytoplankton growth and dynamics (Koseff et al. 1993).

Sandy sediments cover about 50% of the continental shelf seafloor and about 4% of the total ocean floor. Their high permeability allows flow of water through the connected pore network of these sediments. These flows are driven by current or wave induced pressure differences, which may be enhanced through biological structures. Active flows of seawater with suspended particles result in additional transport of particulate organic matter to sediments. Part of the suspended particulate matter entering into these sediments with the water is trapped or utilized by organisms (Huettel et al. 2014), thus supplementing the particulate organic matter supply to these sediments. This additional organic carbon is highly reactive, and particulate organic carbon stocks in permeable sediments are very low (Boudreau et al. 2001).

Another organic carbon provision pathway to sediments involves organisms that consume dissolved organic carbon and use it to form particulate organic carbon (biomass and excretion products), which can be used by other consumers living at the sediment-water interface or within the sediments. This mechanism has been documented for encrusting sponges that consume large quantities of dissolved organic carbon and rapidly convert this into detritus for benthic consumers (i.e. sponge loop, de Goeij et al. 2013).

It is important to realize that some of these additional pathways involve truly new organic carbon (benthic primary production), while the others either convert widely available, but poorly useable, dissolved organic carbon into a particulate pool (sponge loop) or focus particulate organic matter delivered to a wide area for deposition into small areas inhabited by suspension feeding animals or plants. The latter category locally enhances organic matter decomposition at the expense of adjacent areas or water-column degradation. Finally, even in donor-controlled sediments, animals enhance the transfer of labile organic matter to subsurface microbial communities by bioturbation (Box 4.1) and in that way stimulate the overall degradation of organic matter (Middelburg 2018). This mechanism will be presented in detail below; we will first introduce the consuming organisms.

4.2 The Consumers

Organic matter delivered to the sediments represents the major resource for the heterotrophic consumers living in the sediments. These organic carbon consumers are usually partitioned into size classes: microbenthos (<32 μm), meiobenthos (32 μm–0.5/1 mm) and macrobenthos (>0.5/1 mm), the actual size class division depending on the environment (smaller for deep-sea than coastal studies). The microbenthos includes small eukaryotes (Protists), but is by far dominated by prokaryotes (Bacteria and Archaea). The meiobenthos is often dominated by nematodes, but foraminifera (protists) also contribute. The macrobenthos comprises many different animal groups. The number of organisms declines with size: from 10^9 bacteria per ml of sediments (i.e. 10^{14} cells in the top 10 cm of 1 m^2), 10^6 meiofauna m^{-2} to 10^4 macrofauna m^{-2} in estuarine sediments (Herman et al. 1999). The distances between individuals is thus of the order 10 μm, 1 mm and 1 cm for microbes, meiofauna and macrofauna, respectively. Microbes are much smaller than typical pore sizes of silty and sandy sediments, meiofauna can live within these pores, while macrofauna are much larger and consequently move particles when active. The mixing of particles and solutes by animals living in sediment is called bioturbation (Box 4.1) and has major consequences for carbon processing in sediments (Meysman et al. 2006; Middelburg 2018).

   The organic carbon consumed by benthic heterotrophs is used for maintenance, growth and reproduction. The relative importance of the different size classes to carbon consumption scales more or less with their contribution to biomass. In coastal sediments, macrofauna typically accounts for 10–25% of the total biomass and respiration, with microbes accounting for the remaining, except for a small (few %) contribution of protists and meiofauna. Macrofauna number and size (and thus biomass) decline stronglywith water depth compared to those of meiofauna and in particular microbes (Snelgrove et al. 2018). Carbon consumption in deep-sea sediments is almost entirely due to microbes. Accordingly, organic matter consumption is normally attributed to sedimentary microbes, and the consumption of organic matter by animals is therefore ignored in most biogeochemical studies (Middelburg 2018).

4.3   Organic Carbon Degradation in Sediments

Organic carbon degradation in sediments is a highly efficient process because most of the incoming organic carbon is degraded and only a small fraction eventually escapes mineralization to be buried with the accumulating sediments. This fraction is often quantified in terms of the burial efficiency (BE):

$$BE = \frac{F_B}{F_C} = \frac{F_B}{(F_B + R)} \tag{4.1}$$

where F_C is the incoming carbon flux, F_B is the rate of carbon burial and R is the total mineralization rate. Burial efficiencies are often presented as a percentage and range from a fraction of a percent in deep-sea sediments up to tens of a percent in rapidly accumulating coastal sediments (Canfield 1994; Aller 2013). In most oceanic sediments, R >> F_B, and total respiration is a reliable proxy for quantifying the organic carbon delivery rate to sediments (F_C).

This high efficiency of organic matter degradation within in sediments might seem paradoxical because sediments receive organic matter that has already been processed extensively before deposition. Pre-depositional processing of organic matter not only lowers the amount of organic matter delivered, but also the quality because of the preferential use of labile organic carbon by organisms (Fig. 3.9). However, when particulate organic matter is transferred from the water column, with a typical residence time of weeks to months, to the top layer of sediments, the time available for degradation changes by orders of magnitude (10^3–10^4 years). This increase in time available for processing organic matter more than compensates for the loss of reactivity during particle settling; hence, organic matter degradation resumes.

To understand the dynamics and distribution of organic matter in sediments we will again use the simple diffusion-advection-reaction model introduced in Box 1.1. Modelling the distribution and dynamics of organic carbon in sediments is rather challenging because the transport of organic carbon, as well as its production and consumption, are biologically controlled, and we cannot resort to physics-based laws for transport. Moreover, we are dealing with a two-phase problem (particles and water). The first simplification is that we only consider particulate organic carbon. For a sediment with a porosity (volume water/total volume) of 0.7, a particulate organic carbon content of 1 wt% and a very high pore-water dissolved organic carbon concentration (1 mM), dissolved organic carbon contributes about 1‰ to the sediment carbon stock. Moreover, the reactivity of dissolved organic carbon is relative low. Two, gradients in porosity due to compaction and bioturbation are ignored. Three, organic carbon degradation follows first-order kinetics (one-G model, Berner 1964; Box 3.1). Four, electron acceptors are abundant and their nature does not impact organic carbon degradation. Five, solid-phase organic carbon is transported by bioturbation (Box 4.1) and by the net accretion of sediment.

Transport of particulate organic carbon due to animal activity is the dominant transport process in sediment inhabited by animals. The transport resulting from bioturbation can usually be well described by diffusion (Goldberg and Koide 1962; Berner 1980), in particular when the benthic community is diverse (Meysman et al. 2006, 2010). It is only in sediments where chemical conditions (e.g. lack of oxygen, pollution) or physical conditions (e.g. unstable sediments that are repetitively disturbed by waves/tides) restrict macrofauna that bioturbation can be ignored and organic carbon transport due to sediment accretion dominates. Finally by assuming steady-state, the resulting equation for organic carbon (G) is:

$$0 = D_b \frac{d^2 G}{dx^2} - w \frac{dG}{dx} - kG \tag{4.2}$$

where G is labile particulate organic carbon, x is depth (positive downwards), D_b is the bioturbation coefficient ($cm^2 \ yr^{-1}$), w is the sediment accumulation rate (cm y^{-1}) and k is a first-order rate constant (yr^{-1}). Considering that all labile organic carbon is eventually consumed (at depth), we can solve the equation for specific conditions at the sediment-water interface (x = 0). It is instructive to distinguish between sediments that are strictly donor-controlled (i.e. a fixed flux upper boundary) and sediments with a fixed concentration in the top layer. The former applies to most sediments, while the latter would represent coastal sediments with multiple deposition–resuspension events and lateral particulate organic transport pathways (Rice and Rhoads 1989). If we know the concentration of organic carbon at the sediment-water interface (G_0), the solution is (see Box 1.1):

$$G = G_0 e^{\alpha x} \quad \text{with} \quad \alpha = \frac{w - \sqrt{w^2 + 4kD_b}}{2D_b} \tag{4.3}$$

Alternatively, if the flux of organic carbon delivered to the sediment (F) is known the solution is:

$$G = \frac{F}{-D_b \alpha + w} e^{\alpha x}, \tag{4.4}$$

where α is the same as in Eq. 4.3.

Figure 4.2 shows the distribution of organic carbon as a function of sediment depth, for a fixed reactivity and sediment accumulation rate, and for different bioturbation coefficients. The distribution of organic carbon declines exponentially with depth and the attenuation is to a first-approximation governed by $\sqrt{\frac{k}{D_b}}$, i.e. more mixing (higher D_b) will flatten organic carbon concentration versus depth profiles (because the contribution of w to the attenuation coefficient α is very small). In other words, particle mixing by moving animals will transfer organic carbon to larger depths. This is articulated in sediments with a fixed concentration in the top layer. For these sediments bioturbation will increase the total amount of organic

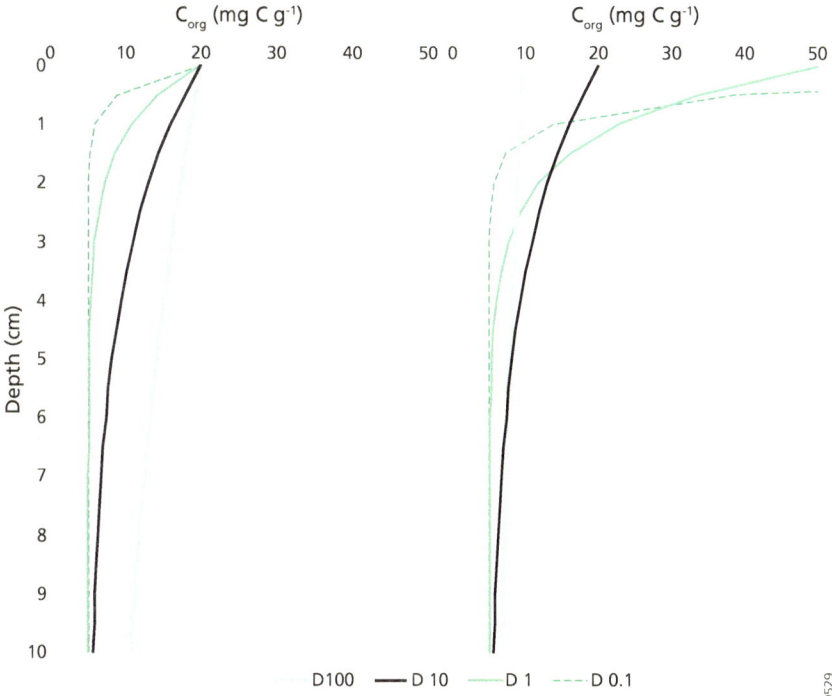

Fig. 4.2 Organic carbon concentration versus depth profiles for different bioturbation coefficients (0.1, 1, 10, 100 cm^2 yr^{-1}) for a fixed concentration of 20 mg C gr^{-1} (left) and a fixed flux of 36 mg C cm^{-2} yr^{-1} (right). The imposed flux has been tuned to generate an identical concentration at the sediment-water interface for a D$_b$ of 10 cm^2 yr^{-1}. Model parameters: k = 1 yr^{-1}; w = 0.1 cm yr^{-1}

matter available for the benthic community. In other words, animals have, to a certain extent, control over their food supply. The other major factor governing organic matter profiles in sediments is the reactivity of organic matter. More reactive organic matter is degraded shallower and does not penetrate as deep into the sediment as refractory organic matter (Fig. 4.3). Moreover, the inventory of organic carbon in surface sediments is higher when the reactivity of organic matter is lower, in particular for a fixed amount of carbon supplied per unit time.

Evidently, when applying this type of models, it is important to select the appropriate upper boundary conditions. While deep-sea sediments can best be described by an organic carbon flux boundary condition, the proper upper boundary condition for coastal sediments is not clear. A fixed concentration upper boundary conditions may be more appropriate for shelf systems in which lateral organic carbon fluxes dominate, e.g. due to a combination of resuspension-deposition cycles and residual tidal flow. Animals collecting organic matter in the surface sediment layer and bioturbating it to deeper layers make thus additional organic matter of high quality available to microbes. Moreover, first-order reactivity

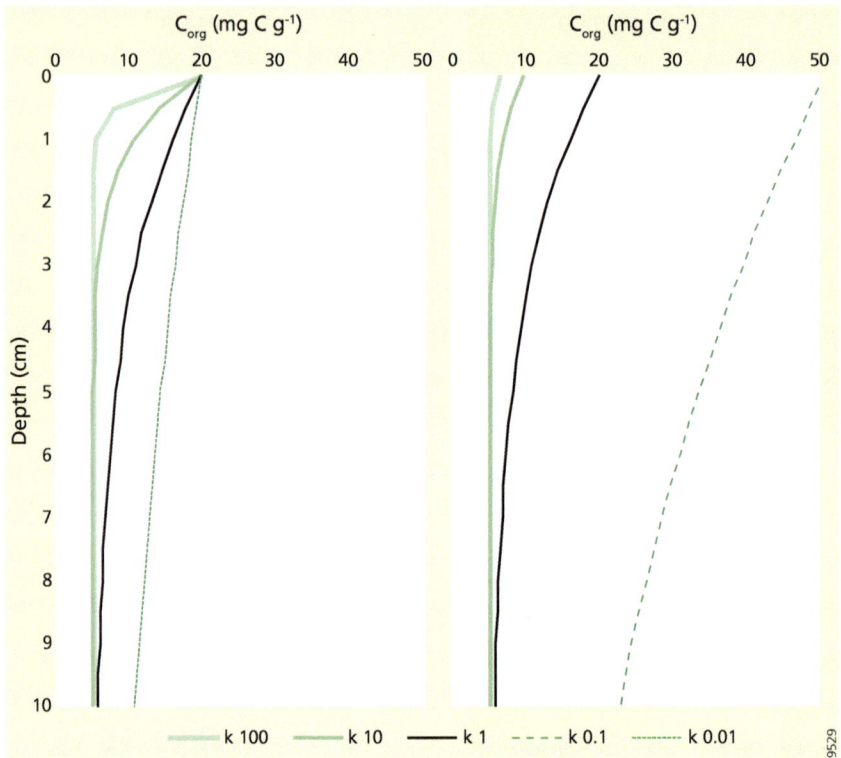

Fig. 4.3 Organic carbon concentration versus depth profiles for different first-order reactivity coefficients (0.1, 1, 10, 100 yr^{-1}) for a fixed concentration of 20 mg C gr^{-1} (left) and a fixed flux of 36 mg C cm^{-2} yr^{-1} (right). The imposed flux has been tuned to generate an identical concentration at the sediment-water interface for k = 1 yr^{-1}. Model parameters: D_b = 10 cm^2 yr^{-1}; w = 0.1 cm yr^{-1}

constants (k) vary over orders of magnitude (Fig. 3.9) and unfortunately cannot be directly measured; they have to derived by fitting a model to observations. This approach is problematic for a number of reasons. One, fitting of a model to observed organic carbon versus depth profiles allows constraining the depth attenuation coefficient $\sqrt{\frac{k}{D_b}}$, but not k or D_b individually (Soetaert et al. 1998; an iconic example of an identifiability problem). Two, both the organic matter reactivity and bioturbation coefficient may depend on depth within sediment. The former because organic matter at depth has on average been exposed longer to degradation and the latter due to the depth distribution of infauna that results in a depth gradient in mixing intensity.

The depth of the bioturbated zone (L), the reactivity of organic carbon (k) and the mixing intensity (D_b) are all related (Boudreau 1998). If we assume that the depth of bioturbation is governed by the availability of food for animals and set a lower limit for organic carbon at depth L (G_L) at 1% of the organic carbon at the surface (G_0):

$$\frac{G_L}{G_0} = 0.01 = e^{\sqrt{\frac{-k}{D_b}}L} \tag{4.5}$$

Making use of $\ln(0.01) \approx 4.6$, this equation can be re-arranged to

$$L = 4.6\sqrt{\frac{D_b}{k}}. \tag{4.6}$$

The depth of the mixed layer is thus directly related the reactivity of the organic matter and the mixing intensity. This equation is highly similar to the more elaborate result of Boudreau (1998) $L = 4\sqrt{\frac{9D_b}{8k}}$ that is based on a resource-feedback model. Figure 4.4 shows the depth of the bioturbated zone as a function of organic matter reactivity and bioturbation intensity. The observed range of Db and k in marine sediments is also indicated. Bioturbation depth in marine sediments vary between 1 and 20 cm, with a global average of about 10 cm (Boudreau 1998). This can be obtained either by high mixing, if organic matter is highly reactive as in coastal sediments, or by low mixing, with less reactive matter in the deep sea.

4.4   Consequences for Sediment Biogeochemistry

Degradation of organic carbon to inorganic carbon requires an electron acceptor, e.g. oxygen. In the water column, oxygen supply is usually sufficient to accept all electron released during respiration, but in sediments solutes such as oxygen are transported primarily by molecular diffusion, and oxygen supply rates often do not match oxygen consumption rates with the consequence that most sediments become anoxic at some depth. In the absence of oxygen, organic matter is either respired anaerobically or fermented. Fermentation refers to an internal redistribution of electrons among carbon atoms with the result that carbon dioxide and methane are formed.

$$2CH_2O \Rightarrow CO_2 + CH_4$$

Anaerobic respiration and fermentation of particulate organic matter involves multiple steps in a complex network: hydrolysis and fermentation of larger molecules to smaller molecules and finally the use of alternative electron acceptors (Jørgensen 2006). These intermediate products include short-chain fatty acids such as formate, acetate, propionate and butyrate, and hydrogen. The latter has a high turn-over (minutes) because of rapid production and consumption, e.g. during hydrogenotrophic processes such as carbon dioxide reduction to methane (e.g. Beulig et al. 2018).

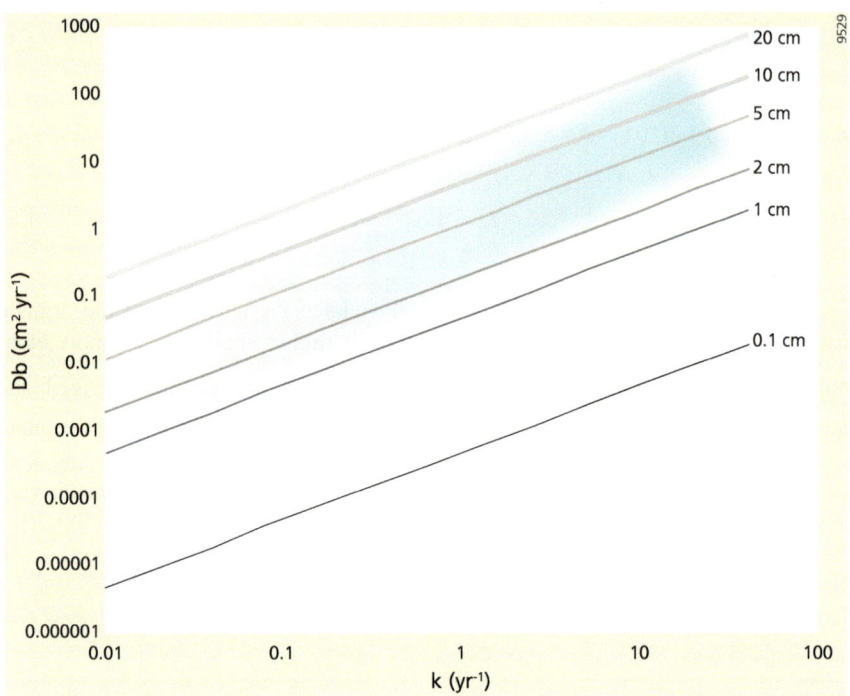

Fig. 4.4 Relationship between organic carbon reactivity (k) and particle mixing (Db) for a number of bioturbation layer depths (0.1, 1, 2, 5, 10 and 20 cm). Sediments receiving reactive organic material are more intensively mixed than those receiving refractory organic matter (blue background box). The global average bioturbation depth is about 10 cm (Boudreau 1994)

The most important alternative terminal electron acceptors are nitrate and nitrite, manganese and iron oxides and hydroxides, and sulphate, and they are used in this sequence largely because of multiple factors, including energy yield. Table 4.1 provides the overall reaction for organic carbon respiration based on nitrate, manganese oxide, iron oxide and sulphate and the corresponding change in Gibbs free energy.

The Gibbs free energy change declines from aerobic respiration to denitrification (nitrate reduction to dinitrogen gas) and metal oxide reduction to sulfate reduction because of the formation of reduced compounds (metabolites) such as ammonium, manganese (II), iron (II), hydrogen sulphide and methane (Fig. 4.5). These reduced metabolites contain a substantial amount of energy that was originally part of the organic matter. Once produced these reduced metabolites can react (after diffusion upward) with an electron acceptor (higher in the redox cascade). This re-oxidation of reduced metabolites guarantees that sediments efficiently utilize all organic matter energy delivered (Fig. 4.5). For instance, methane generated at depth can be oxidized anaerobically with sulphate, metal oxides and nitrate/nitrite and aerobically. Some of the metabolites generated (e.g. sulphide) may react with sedimentary iron containing minerals. The iron sulphide formed may be buried, but the majority

Table 4.1 Standard gibbs free energy change of organic matter degradation pathways (based on Berner 1980)

Degradation pathway	Reaction	ΔG^0 (kJ mol^{-1} of CH$_2$O)
Aerobic respiration	$CH_2O + O_2 \Leftrightarrow CO_2 + H_2O$	−475
Denitrification	$CH_2O + 0.8\ NO_3^- \Leftrightarrow 0.4\ N_2 + 0.8\ HCO_3^- + 0.2\ CO_2 + 0.6\ H_2O$	−448
Manganese oxide reduction	$CH_2O + 3\ CO_2 + 2\ MnO_2 + H_2O \Leftrightarrow 2\ Mn^{2+} + 4\ HCO_3^-$	−349
Iron(hydr)oxide reduction	$CH_2O + 7\ CO_2 + 4\ Fe(OH)_3 \Leftrightarrow 4\ Fe^{2+} + 8\ HCO_3^- + 3\ H_2O$	−114
Sulfate reduction	$CH_2O + 0.5\ SO_4^{2-} \Leftrightarrow 0.5\ H_2S + HCO_3^-$	−77
Methanogenesis	$CH_2O \Leftrightarrow 0.5\ CO_2 + 0.5\ CH_4$	−58

The decrease in Gibbs free energy of the reaction results from the accumulation of energy in the reaction products, i.e. the metabolites

is usually re-oxidized after bioturbation to surface layers where oxygen or nitrate are present (Jørgensen 1977; Berner and Westrich 1985). Oxidation of these reduced metabolites often involves multiple reactions: e.g. hydrogen sulphide is oxidized to intermediate compounds such as elemental sulfur, thiosulfate and sulphite, which may then be oxidized further or disproportionated (i.e. splitted into a more reduced and more oxidized compound).

This efficient re-oxidation of reduced metabolites has two important consequences for studying the sedimentary carbon cycle. First, the energy released during re-oxidation reactions can be used to fix inorganic carbon, i.e. it supports chemolithoautotrophic processes in sediments. Chemolithoautotrophy in marine sediments is still poorly constrained, but likely on the order of 0.3–0.4 Gt C yr^{-1} (Middelburg 2011). Secondly, total oxygen consumption is not only linked to aerobic respiration, but also includes oxygen consumption during re-oxidation processes. In organic-carbon limited sediments in the deep-sea, oxygen supply by diffusion is sufficient and all organic matter mineralization occurs aerobically, and oxygen use is primarily due to oxidation of organic carbon to carbon dioxide and nitrification (oxidation of ammonium to nitrate). In sediment with high carbon loadings, most of the oxygen consumption is due to the oxidation of reduced sulfur, iron, manganese and methane, and total oxygen consumption provides an excellent proxy for total sediment respiration irrespective of actual respiration occurred aerobically or anaerobically (Fig. 4.6). In fact, total respiration of sediments is equal to total oxygen consumption minus the burial of reduced iron and sulphide and the escape of nitrogen gas over the sediment-water interface. These contributions are usually limited to a few percent and oxygen consumption can, therefore, be used as a measurable proxy for total respiration without the need to study all microbiological reduction and oxidation processes in detail.

Oxygen distributions in sediments are normally modelled with a simple model balancing the consumption of oxygen with the diffusive supply:

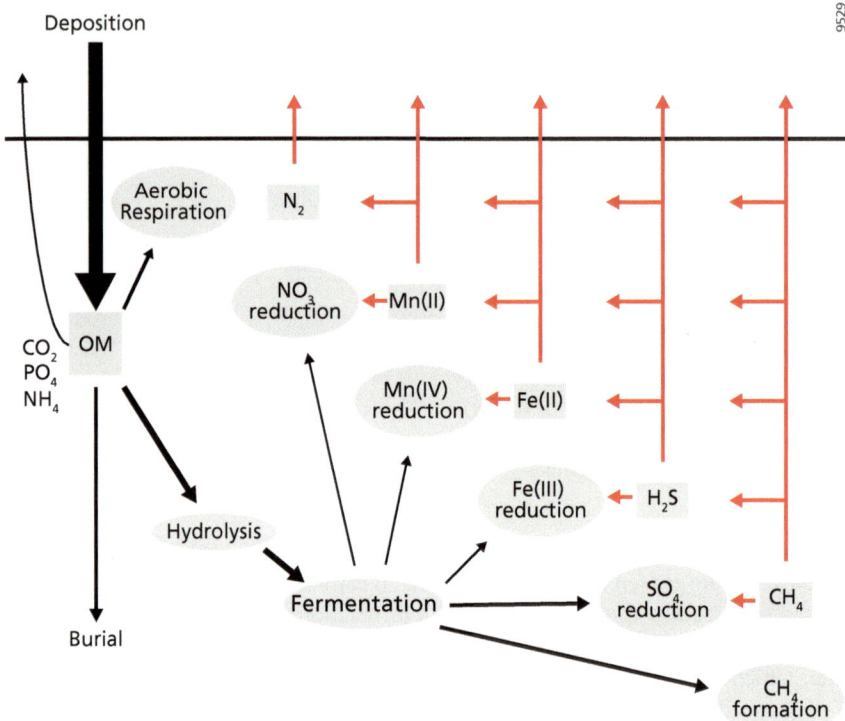

Fig. 4.5 Conceptual model of organic matter (OM) degradation and reoxidation pathways (based on Jørgensen 2006, and Middelburg and Levin 2009). The red arrows reflect the fate of (energy-rich) substrates released during anaerobic mineralization

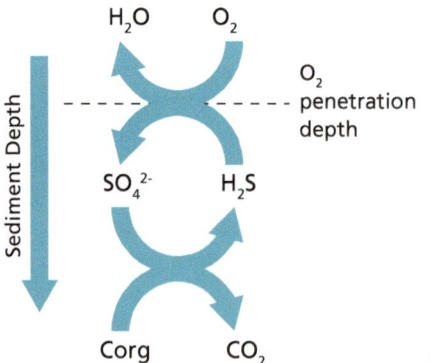

Fig. 4.6 Conceptual model of tight coupling sulfate utilization for respiration of organic carbon and sulphide oxidation by oxygen, with the net result that one mole of oxygen use corresponds with one mole of organic carbon oxidized to carbon dioxide (after Aller 1994)

$$D_s \frac{d^2 O_2}{dx^2} = R_0 \tag{4.7}$$

where O_2 is the oxygen concentration, D_s is the diffusion coefficient of oxygen in sediments and R_0 is the consumption of oxygen. Oxygen consumption rates can be described with a depth and oxygen concentration independent rate because re-oxidation is the dominant oxygen sink. The solution for this zero-order kinetics equation with a fixed concentration at the sediment-water interface (C_0) is (Bouldin 1968):

$$C = \frac{R_0}{2D_s} x^2 - \left(\sqrt{\frac{2C_0 R_0}{D_s}} \right) x + C_0 = C_0 \left(1 - \frac{x}{\delta} \right)^2 \tag{4.8}$$

where δ is the oxygen penetration depth:

$$\delta = \sqrt{\frac{2C_0 D_s}{R_0}} \tag{4.9}$$

This relation shows that the oxygen penetration depth is inversely related to the total respiration. High respiration rates in coastal sediment result in shallow oxygen penetration (mm scale), while low respiration rates in organic carbon starved deep-sea sediments allow oxygen penetration to meters (Fig. 4.7). Moreover, oxygen penetration depth exhibits a square-root dependence on bottom-water oxygen levels (C_0).

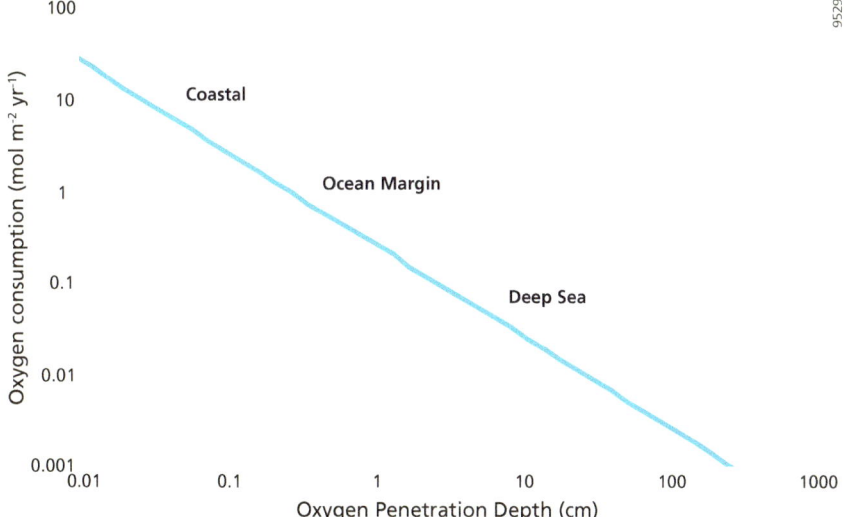

Fig. 4.7 The relationship between oxygen consumption and oxygen penetration depth. High fluxes of oxygen into sediments imply shallow penetration. Oxygen penetration varies from <1 mm in coastal sediments to >1 m in deep-sea sediment receiving low carbon loadings

4.5 Factors Governing Organic Carbon Burial

Following active carbon degradation in the water column and continued mineralization in surface sediments, at some depth in the sediment all labile organic matter has been utilized and a background level of organic carbon remains (Figs. 4.2 and 4.3). This organic carbon is buried down with the accretion of sediments. Organic carbon burial is, thus, the product of sediment accumulation and organic carbon content:

$$F_B = w(1 - \phi)\rho_s G \qquad\qquad (4.10)$$

where w is the sediment accumulation rate (cm yr^{-1}), ϕ is the porosity, ρ_s is the dry density (gr cm^{-3}), G is the concentration of organic carbon (g C gr^{-1}), with F_B expressed in g C cm^{-2} yr^{-1}. It is important to realize that w refers to the long-term accumulation rate at depth and not the rate of sediment deposition at the surface. The concentration of organic carbon should theoretically be at the depth where no further degradation occurs. Such a depth would not exist if organic matter decay follows a reactive continuum: organic matter is reactive on all timescale. However, in most sediments organic carbon concentration rapidly declines with depth and then remains almost invariant. The precise reference depth is not so important as long the gradient can be neglected.

Differences in organic carbon burial are primarily determined by differences in sediment accumulation and carbon contents, and only secondarily by differences in porosity and density. Sediment porosities vary 0.4–0.8 and dry sediment densities from 2.6 gr cm^{-3} for mineral to ~ 2.0 gr cm^{-3} for peaty sediment. Organic carbon concentrations depend on the carbon input to sediments and the extent of decay before burial. Any process that enhances organic carbon input or impedes degradation elevates organic carbon concentration at depth and consequently carbon burial. These two factors relate directly to the old discussion on whether high export production (carbon delivery to sediment) or an oxygen related effect on preservation governs burial of organic carbon. The idea is that organic matter degradation in the presence of oxygen is more efficient than without (Box 4.2).

The by far most important factor governing organic carbon burial is the delivery of inorganic particles that results in sediment accretion. Sediment accumulation in the ocean vary about four to five orders of magnitude, from less than 0.1 cm per thousand year in the deep-sea to centimetres per year in rapidly accumulating coastal sediments (Aller 2013). Most organic carbon burial, therefore, occurs in coastal sediments. An increase in inorganic particle deposition should, in principle, not increase carbon burial, because the organic carbon will just be diluted, but observations show otherwise. In nature, particulate organic carbon is mainly associated with mineral surfaces and the higher the accumulation of minerals, the more mineral surfaces, with associated organic matter, are buried. Vegetated coastal sediments are another site of high organic carbon burial for two reasons. First, the presence of macrophytes stabilizes sediment and enhances deposition via canopy trapping of particles. Secondly, the net ecosystem carbon balance of these communities results in the accumulation of plant derived organic matter. The following flux balance can be used to investigate the impact of sediment accretion and net ecosystem production on the burial of organic carbon in vegetated sediments:

$$w(1 - \phi)\rho_s C_{external} + \rho_{organic} CtoM \, NEP = w(1 - \phi)\rho_s C_{buried} \tag{4.11}$$

where w is sediment accumulation rate (cm yr^{-1}), ϕ is the porosity, ρ_s is the dry density of bulk sediment (gr cm^{-3}), $C_{external}$ is the concentration of organic carbon (g C gr^{-1}) in deposited sediments remaining after degradation, $\rho_{organic}$ is the density of plant organic material, CtoM converts organic C to organic matter, and C_{buried} is the concentration of buried organic matter. Primary production in cordgrass habitats is very high (~ 2 kg C m^{-2} yr^{-1}) with net carbon accretion of ~ 0.3 kg C m^{-2} yr^{-1} (Middelburg et al. 1997). If all that cordgrass detritus is retained in the system, marsh sediment accretion would be ~ 0.5 mm per year, not enough for keeping pace with accelerated global sea-level rise (>2 mm per year) and a mineral starved, peaty marsh may therefore drown. The combined effect of enhanced particle trapping and retention and macrophyte derived organic matter are the reason that vegetated coastal sediments represent a major term in global marine carbon burial, i.e. blue carbon (Duarte 2017).

Box 4.1: Bioturbation

Bioturbation refers to the reworking of sediments (particles and water) due the activities of organisms, in particular animals (Meysman et al. 2006). This reworking has many consequences, including the creation of a three-dimensional, heterogeneous distribution of sediment properties. In fact, this animal induced heterogeneity is clear from sediments deposited during the last 0.5 billion years and palaeontologists have therefore named this period Phanerozoic, after the old greek words phanerós (visible) and zōé (life). Moreover, the small-scale heterogeneity stimulates biogeochemical and biological diversity. It also has consequences for porosity, permeability, sediment stability and roughness.

Besides these effects on sediment properties and characteristics, animal activities have a major impact on the transport of particles and solutes. Particle mixing, bioturbation sensu stricto, occurs through the construction and maintenance of structures such as burrows and the movement of particles while feeding (ingestion and defecation). Particles are moved vertically and horizontally, but the consequences are usually more prominent vertically because most sediment characteristics show strong vertical gradients. The particles dispersed may be reactive or unreactive minerals, organic matter as well as organisms and their cysts, eggs and remains. Many animals construct and live in burrows and ventilate these with overlying oxygenated water for respiration or food supply. In permeable sediments, this burrow water can enter the sediments and in permeable and non-permeable sediments diffusive exchange occurs between burrows and adjacent sediments because of concentration gradients. Many animals are involved in particle mixing and ventilation. Particle mixing is pivotal for optimal functioning of sediments: organic matter and solid-phase oxidants such as iron oxides are mixed downwards, while reduced substance such as iron sulphides are mixed upwards. Bio-irrigation enhances exchange of oxygen, nutrients and other substances between water and sediment.

Particle mixing is the result of multiple activities of a diverse assemblage of large animals. While the behaviour of certain species may cause a distinct particle transport pattern, the net result of many particle mixing events can normally be described very well as diffusion. Bioturbation is then quantified in terms of a biodiffusion coefficient for particles (D_b; cm^2 yr^{-1}). This biodiffusion coefficient can be derived from modelling concentration versus depth profiles of a tracer A with a known reactivity (e.g. radioactive decay). Equation 4.12 balances transport due to sediment accumulation and biodiffusion with radioactive decay at steady state:

$$0 = D_b \frac{d^2A}{dx^2} - w\frac{dA}{dx} - kA \qquad (4.12)$$

where w is the sediment accumulation rate (cm yr^{-1}) and k is the radioactive decay constant (yr^{-1}). These macroscopic empirical D_b values can be decomposed into discrete particle properties step length (λ) and the period between two displacements (τ) (Meysman et al. 2010):

$$D_b = \frac{\lambda^2}{2\tau} \qquad (4.13)$$

The step length of particle displacement scales with organism size, and it is for this reason that meiofauna and similarly sized organisms contribute little to particle displacement.

Box 4.2: Oxygen and Organic Matter Preservation

The impact of oxygen on organic matter degradation and preservation has been studied extensively because of its relevance for the formation of oil and gas source rocks, the reconstruction of paleoenvironments, and the projection of carbon cycling in the future warmer, low oxygen ocean. Although most of the evidence is correlative, there is now consensus that more organic carbon is preserved under low oxygen conditions (Middelburg and Levin 2009). Experimental studies have shown that oxygen has little if any impact on microbial organic matter degradation, but is needed for the degradation of the most refractory fraction of sediment organic carbon (Hulthe et al. 1998). Natural experiments provided by turbidite deposition and cyclic de-oxygenation events have revealed that organic matter preserved under anoxic conditions can be re-activated when exposed to dissolved oxygen (Moodley et al. 2005). Bottom-water oxygen levels have been shown to increase the carbon burial efficiency and the amount of carbon preserved per unit area of reactive surface area (Canfield 1994; Hartnett et al. 1998). Lack of oxygen not only increases the quantity of organic carbon buried, but also its nature. Organic carbon buried under anoxic conditions is usually less degraded and that is reflected in organic matter proxies (see Chap. 6), such as hydrogen index and the amino-acid degradation index (Middelburg and Levin 2009). Elevated organic carbon burial under anoxic conditions is a negative feedback in Earth System dynamics.

Although the impact of dissolved oxygen on carbon preservation has been well documented, and we have made progress studying the implications, there is still little understanding of why there is more organic carbon burial under anoxic conditions. Most organic matter is buried in ocean margin and coastal sediments that are anoxic below a few mm to cm, irrespective of the presence of oxygen in the bottom waters. This implies that the dissolved oxygen effect should be either pre-depositional or related to the changes in the benthic community processing the delivered organic matter. There is evidence for both. Organic carbon flux attenuation in an anoxic water column is less than that in an oxic water column, perhaps due to the absence of zooplankton under anoxic conditions, and the organic matter delivered to anoxic sediments appears to be less reactive towards degradation (Keil et al. 2016). Changes in bottom-water oxygen levels have consequences for the benthic community composition because of the decrease or disappearance of animals (Jessen et al. 2017). Benthic animals play a major role in sediment ecosystem functioning, through their interactions with microbes: particle mixing delivers labile organic carbon to microbes living at depth and bio-irrigation provides microbes with fluctuating oxygen levels, so that they can efficiently process organic matter (Aller 2013).

References

Aller RC (1994) Bioturbation and remineralization of sedimentary organic matter—effects of redox oscillation. Chem Geol 114:331–345

Aller RC (2013) Sedimentary diagenesis, depositional environments, and benthic fluxes treatise on geochemistry, vol 8, 2nd edn. Elsevier, Oxford, pp 293–334

Barrón C, Middelburg JJ, Duarte CM (2006) Phytoplankton trapped within seagrass (*Posidonica oceanica*) sediments is a nitrogen source: an in situ isotope labeling study. Limnol Oceanogr 51:1648–1653

Berner RA (1964) An idealized model of dissolved sulfate distribution in recent sediments. Geochimica et Cosmochimica Acta 28:1497–1503

Berner RA (1980) Early diagenesis: a theoretical approach. Princeton University Press, Princeton

Berner RA, Westrich JT (1985) Bioturbation and the early diagenesis of carbon and sulphur. Am J Sci 285:193–206

Beulig F, Røy H, Glombitza C, Jørgensen BB (2018) Control on rate and pathway of anaerobic organic carbon degradation in the seabed. Proc Natl Acad Sci 15:367–372

Boudreau BP (1994) Is burial velocity a master parameter for bioturbation? Geochim Cosmochim Acta 59:1243–1249

Boudreau BP (1998) Mean mixed depth of sediments: the wherefore and the why. Limnol Ocean 3. https://doi.org/10.4319/lo.1998.43.3.0524

Boudreau BP, Huettel M, Forster S, Jahnke RA, McLachlan A, Middelburg JJ, Nielsen P, Sansone F, Taghon G, van Raaphorst W, Webster I, Weslawski JM, Wiberg P, Sundby B (2001) Permeable marine sediments: overturning an old paradigm. EOS 82(11):133–136

Bouldin DR (1968) Models for describing diffusion of oxygen and other mobile constituents across mud-water interface. J Ecol 56:77–87

Canfield DE (1994) Factors influencing organic carbon preservation in marine sediments. Chem Geol 114:315–329

de Goeij JM, van Oevelen D, Vermeij MJA, Osinga R, Middelburg JJ, de Goeij AFP, Admiraal W (2013) Surviving in a marine desert: the sponge loop retains resources within coral reefs. Science 342:108–110

Duarte CM (2017) Reviews and syntheses: hidden forests, the role of vegetated coastal habitats in the ocean carbon budget. Biogeosciences 14:301–310

Duarte CM, Cebrian J (1996) The fate of marine autotrophic production. Limnol Oceanogr 41:1758–1766

Gacia E, Duarte CM, Middelburg JJ (2002) Carbon and nutrient deposition in a Mediterranean seagrass (Posidonia oceanica) meadow. Limnol Oceanogr 47:23–32

Gattuso JP, Gentilli B, Duarte C, Kleypass J, Middelburg JJ, Antoine D (2006) Light availability in the coastal ocean: impact on the distribution of benthic photosynthetic organism and contribution to primary production. Biogeosciences 3:489–513

Goldberg ED, Koide M (1962) Geochronological studies of deep-sea sediments by the Io/Th method. Geochim Cosmochim Acta 26:417–450

Hartnett HE, Keil RG, Hedges JI, Devol AH (1998) Influence of oxygen exposure time on organic carbon preservation in continental margin sediments. Nature 391:572–574

Heip CHR, Goosen NK, Herman PMJ, Kromkamp J, Middelburg JJ, Soetaert K (1995) Production and consumption of biological particles in temperate tidal estuaries. Oceanogr Mar Biol Ann Rev 33:1–150

Herman PMJ, Middelburg JJ, van de Koppel J, Heip CHR (1999) Ecology of estuarine macrobenthos. Adv Ecol Res 29:195–240

Huettel M, Berg P, Kostka JE (2014) Benthic exchange and biogeochemical cycling in permeable sediments. Ann Rev Mar Sci 6:23–51

Hulthe G, Hulth S, Hall POJ (1998) Effect of oxygen on degradation rate of refractory and labile organic matter in continental margin sediments. Geochim Cosmochim Ac 62:1319–1328

Jessen GL, Lichtschlag A, Ramette A, Pantoja S, Rossel PE, Schubert CJ, Struck U, Boetius A (2017) Hypoxia causes preservation of labile organic matter and changes microbial community composition (Black Sea shelf). Sci Adv 3:e1601897. https://doi.org/10.1126/sciadv.1601897

Jørgensen BB (1977) Sulphur cycle of a coastal marine sediment (Limfjorden, Denmark). Limnol Oceanogr 22:814–832

Jørgensen BB (2006) Bacteria and marine biogeochemistry. In: Shulz HD, Zabel M (eds) Marine geochemistry. Springer, Berlin, pp 169–206

Keil RG, Neibauer JA, Biladeau C, van der Elst K, Devol AH (2016) A multiproxy approach to understanding the "enhanced" flux of organic matter through the oxygen-deficient waters of the Arabian Sea. Biogeosciences 13:2077–2092

Kennedy H, Beggins J, Duarte CM, Fourqurean JW, Holmer M, Marba N, Middelburg JJ (2010) Seagrass sediments as a global carbon sink: isotope constraints. Global Biogeochem Cycles. https://doi.org/10.1029/2010gb003848

Koseff JR, Holen JK, Monismith SG, Cloern JE (1993) Coupled effects of vertical mixing and benthic grazing on phytoplankton populations in shallow, turbid estuaries. J Mar Res 51:843–868

Meysman FJR, Middelburg JJ, Heip CHR (2006) Bioturbation: a fresh look at Darwin's last idea. Trends Ecol Evol 21:688–695

Meysman FJR, Boudreau BP, Middelburg JJ (2010) When and why does bioturbation lead to diffusive mixing? J Mar Res 68:881–920

Middelburg JJ (2011) Chemoautotrophy in the ocean Geoph. Res Lett 38:L24604. https://doi.org/10.1029/2011GL049725

Middelburg JJ (2018) Reviews and syntheses: to the bottom of carbon processing at the seafloor. Biogeosciences 5:413–427

Middelburg JJ, Levin LA (2009) Coastal hypoxia and sediment biogeochemistry. Biogeosciences 6:1273–1293

Middelburg JJ, Nieuwenhuize J, Lubberts RK, van de Plassche O (1997) Organic carbon isotope systematics of coastal marshes. Estuar, Coast Shelf Sci 45:681–687

Middelburg JJ, Barranguet C, Boschker HTS, Herman PMJ, Moens T, Heip CHR (2000) The fate of intertidal microphytobenthos carbon: an in situ ^{13}C labelling study. Limnol Oceanogr 45:1224–1234

Moodley L, Middelburg JJ, Herman PMJ, Soetaert K, de Lange GJ (2005) Oxygenation and organic-matter preservation in marine sediments: direct experimental evidence from ancient organic carbon-rich deposits. Geology 33:889–892

Rice DL, Rhoads DC (1989) Early diagenesis of organic matter and the nutritional value of sediment. In: Lopez G, Taghon G, Levinton J (eds) Ecology of marine deposit feeders. Springer, Berlin, pp 309–317

Snelgrove PV, Soetaert K, Solan M, Thrush S, Wei CL, Danovaro R, Fulweiler RW, Kitazato H, Ingole B, Norkko A, Parkes RJ (2018) Global carbon cycling on a heterogeneous seafloor. Trends Ecol Evol 33:95–105

Soetaert K, Herman PMJ, Middelburg JJ, Heip CHR (1998) Assessing organic matter mineralization, degradability and mixing rate in an ocean margin sediment (Northeast Atlantic) by diagenetic modelling. J Mar Res 56:519–534

Biogeochemical Processes and Inorganic Carbon Dynamics

5

Organic matter production and degradation result in the consumption and release of carbon dioxide; oxidation and reduction reactions involve the production or consumption of protons; carbonate mineral formation and dissolution reactions cause consumption or release of carbonate and bicarbonate. All these processes have the potential to increase or decrease proton concentrations, i.e. change the pH, yet marine pH varies over a rather small range (about one unit only). This is due to the buffering of seawater. For a complete understanding of the impact of carbon dioxide on organisms, the role of organisms in carbon dioxide dynamics, and the role of marine systems in climate change (e.g., carbon dioxide uptake and ocean acidification; Box 5.1), we have to understand the chemistry of the carbon dioxide system in water and for this we need to understand what governs pH dynamics.

In this chapter, we will first refresh the basics of inorganic carbon chemistry in water and seawater, then introduce alkalinity and buffering, and discuss how biological processes and mineral formation are impacted and impact carbon dioxide in marine systems.

5.1 The Basics

Pure water can dissociate into protons and hydroxide ions:

$$H_2O \Leftrightarrow H^+ + OH^- \tag{5.1}$$

This reaction occurs virtually immediately and one can thus assume equilibrium between the three species ($[H_2O]$, $[H^+]$, $[OH^-]$):

$$K = \frac{[H^+][OH^-]}{[H_2O]} \quad \text{or} \quad K_w = [H^+][OH^-] \tag{5.2}$$

© The Author(s) 2019
J. J. Middelburg, *Marine Carbon Biogeochemistry*, SpringerBriefs in Earth System Sciences, https://doi.org/10.1007/978-3-030-10822-9_5

where K_w denotes the equilibrium constant for water self-ionisation. This type of equilibrium relation is also known as a mass-action law. Having two unknowns ($[H^+]$, $[OH^-]$) that are related via the above equilibrium relation, we have one degree of freedom: i.e. if we choose $[H^+]$, then $[OH^-]$ is set to $\frac{K_w}{[H^+]}$.

Next, we consider pure water in equilibrium with a gas containing CO_2. Part of the CO_2 will remain in the atmosphere ($CO_{2(g)}$), but part will dissolve in water ($CO_{2(aq)}$). This dissolved $CO_{2(aq)}$ will react with water to form carbonic acid (H_2CO_3) according to:

$$H_2O + CO_{2(aq)} \Leftrightarrow H_2CO_3 \tag{5.3}$$

This equilibrium is somewhat slow and the equilibrium constant of hydration $K_h = \frac{[H_2CO_3]}{[H_2O][CO_2]}$ is rather small (~ 0.002); almost all dissolved carbon dioxide remains in the form of $CO_{2(aq)}$. Moreover, it is analytically not possible to distinguish between $CO_{2(aq)}$ and $H_2CO_{3(aq)}$; they are therefore lumped together and usually termed CO_2^* or $H_2CO_3^*$. However, carbonic acid $H_2CO_{3(aq)}$ is a moderately weak acid ($K \approx 10^{-3.6}$), while the combined $H_2CO_3^*$ is a weak acid ($K \approx 10^{-6.3}$). For reasons of notational simplicity, we will term it H_2CO_3 from now onwards.

Carbonic acid is a weak diprotic acid and partly dissociates first into a bicarbonate ion (HCO_3^-) and a proton, and subsequently, the bicarbonate is dissociated partly into carbonate ion (CO_3^{2-}) and a proton. The relevant reactions are:

$$H_2CO_3 \Leftrightarrow HCO_3^- + H^+ \tag{5.4}$$

$$HCO_3^- \Leftrightarrow CO_3^{2-} + H^+ \tag{5.5}$$

for which we can write equilibrium relations:

$$K_1 = \frac{[HCO_3^-][H^+]}{[H_2CO_3]} \tag{5.6}$$

and

$$K_2 = \frac{[CO_3^{2-}][H^+]}{[HCO_3^-]} \tag{5.7}$$

where K_1 and K_2 are the first and second equilibrium constants ($10^{-6.35}$ and $10^{-10.3}$ in freshwater at 25 °C). The relative concentrations of $[H_2CO_3]$, $[HCO_3^-]$ and $[CO_3^{2-}]$ are governed by the pH ($-\log_{10}[H^+]$) of the solution as depicted by a Bjerrum plot (Fig. 5.1). Carbonic acid is the dominant species at pH values below the pK_1 ($-\log_{10}K_1$), bicarbonate dominates between the pK_1 and pK_2 values and the carbonate ion dominates at pH values above the pK_2.

The carbonic acid in solution and carbon dioxide in the atmosphere are related via Henry's law:

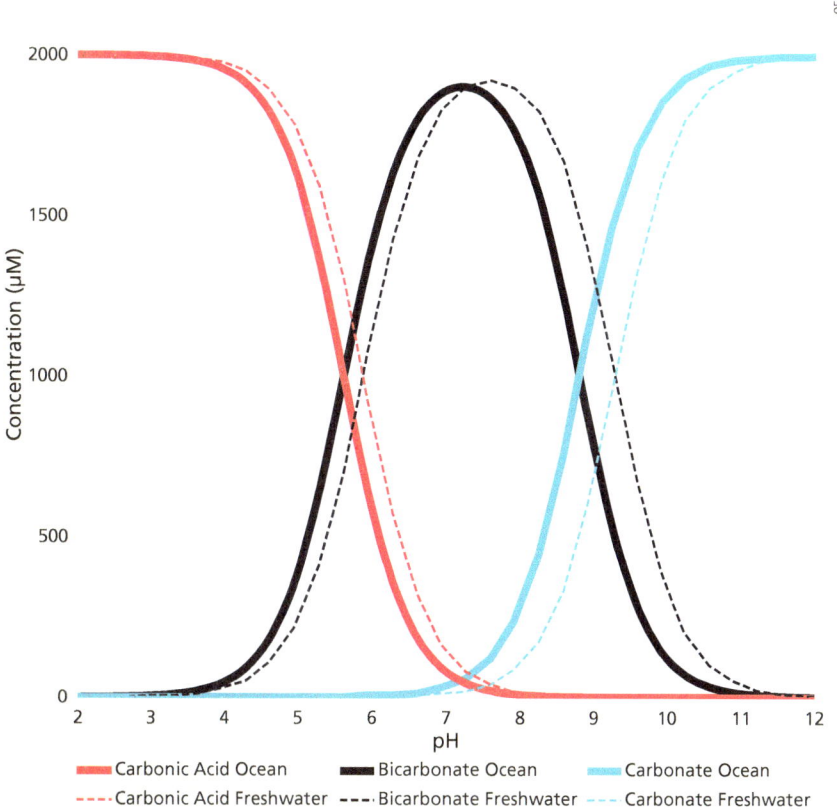

Fig. 5.1 Bjerrum plot showing the distribution of carbonic acid (red), bicarbonate (black) and carbonate (blue) ions as a function of pH in freshwater (dashed line) and seawater (solid lines). DIC = 2000 μM

$$[H_2CO_3] = K_H \times pCO_2 \tag{5.8}$$

where K_H is the Henry constant for CO_2 solubility in water (mol kg^{-1} atm^{-1}) and pCO_2 is the partial pressure of CO_2 (atm). The equilibrium constants K_w, K_1, K_2 and K_H are functions of temperature, pressure and salinity in seawater.

Accordingly, for the CO_2–H_2O system, we have five unknown concentrations ([H_2CO_3], [HCO_3^-], [CO_3^{2-}], [OH^-] and [H^+]) and four equilibrium relations: water self-ionisation (5.2), Henry's law (5.8) and the first and second equilibria of carbonic acid dissociation (5.6, 5.7). To solve the system, we need one additional relation. Natural waters are uncharged, and we can thus use the charge balance equation for this system: the positive charge of protons should be balanced by the negative charge of hydroxide, bicarbonate and carbonate ions.

$$[H^+] = [OH^-] + [HCO_3^-] + 2[CO_3^{2-}] \tag{5.9}$$

Note that the carbonate ion is counted twice in a charge balance because of its double charge. Alternatively, one can define a proton balance equation, a mass balance for protons (Butler 1982):

$$[H^+] = [H^+]_{H_2O} + [H^+]_{H_2CO_3} \qquad (5.10a)$$

or its equivalent

$$[H^+] = [OH^-] + [HCO_3^-] + 2[CO_3^{2-}], \qquad (5.10b)$$

This proton conservation equation balances excess protons on the left-hand side with the recipe on the right-hand side. Proton mass balances are always relative to a proton reference level, e.g. Equation (5.10a, b) is relative to H_2CO_3 and in this case, it is identical to the charge balance. Mathematically, the system is now fully defined with 5 unknown species linked via 5 equations. Moreover, the system remains fully determined if another component is added for which the total concentration and equilibrium distribution among species is known.

5.2 The Thermodynamic Basis

So far, we have ignored non-ideal behaviour of gases, liquids and solutes that have an impact on equilibria. Although the name equilibrium constant suggests that it is a constant, it is in fact not, because of its dependence on temperature, pressure and the composition of the solution. This dependence relates to the effect of temperature, pressure and mixture composition on the Gibbs Free energy. Formally, at equilibrium, the change in Gibbs free energy ΔG^{react} (J mol^{-1}) is related to the thermodynamic equilibrium constant (K_x^{therm}) via $\Delta G^{react} = -RT \ln K_x^{therm}$, where T is in Kelvin and R is the universal gas constant (8.314 J mol^{-1} K^{-1}).

There are two generic pathways towards obtaining equilibrium constants for complex natural solutions. The first approach involves calculating thermodynamic constants from the well-documented and tabulated standard Gibbs free energies of formation and correcting these for temperature and pressure of the system. The thermodynamic equilibrium constant is then expressed in activities of the molecules involved, rather than concentrations. For example, equilibrium relation (5.3) for the thermodynamic K_1^{therm} should formally be written as:

$$K_1^{therm} = \frac{\{HCO_3^-\}\{H^+\}}{\{H_2CO_3\}}$$

where the {} indicate that activities, rather than concentrations [] are used. Concentrations and activities are linked via activity coefficients (γ), which account for both electrostatic interactions, as well as formation of ion-pairs among the various ions in solution:

$$K_1^{therm} = \frac{\{HCO_3^-\}\{H^+\}}{\{H_2CO_3\}} = \frac{[HCO_3^-][H^+]}{[H_2CO_3]} \times \frac{\gamma_{HCO_3}\,\gamma_{H^+}}{\gamma_{H_2CO_3}}$$

These activity coefficients are usually calculated using the Pitzer model (Millero 2007) that can be applied to highly complex media, such as brines.

The alternative way of estimating equilibrium constants, used by most chemical oceanographers and marine biogeochemists, is to experimentally determine stoichiometric constants as a function of temperature, pressure and salinity and represent this dependency by a polynomial function. The underlying idea is that the relative composition of seawater is rather constant, that the reaction as such does not change the composition of seawater and that direct measurements in seawater provide more accurate results than using thermodynamic data with activity coefficient corrections:

$$K_{1,(T,S,P)}^* = \frac{[HCO_3^-][H^+]}{[H_2CO_3]}$$

where the * superscripts indicates that it is a stoichiometric quantity, rather than thermodynamic constant and the subscripts T, S and P stand for temperature, salinity and pressure. The temperature, salinity and pressure dependence of stoichiometric constants result in rather complex and cumbersome expressions (Table 5.1), but there are many computer programs available to facilitate their determination.

Figure 5.2 shows the stoichiometric constant expressed as pK_1^* as a function of temperature and salinity. pK_1^* values decrease with increasing temperature and salinity, or in other words, carbonic acid is more dissociated with increasing temperature or salinity. The ratio of the stoichiometric equilibrium constant (K_1^*) at salinity 35 to the thermodynamic equilibrium constant (K_1^{therm}) in pure water is about 4.4–5. This salinity effect has consequences for the speciation of dissolved inorganic carbon in seawater (Fig. 5.1). Similar dependencies apply to other stoichiometric constants. Moreover, this stoichiometric approach can also be used to accurately quantify the solubility of minerals in seawater and will be used from now on (while dropping the * superscript).

The solubility of gases, including CO_2, is also a function of temperature, salinity and pressure (increasing salinity and temperature lower the solubility of gases). Thermodynamically, the fugacity of carbon dioxide is linked via the Henry constant to dissolved carbon dioxide: $[H_2CO_3] = K_H \times fCO_2$. The fugacity is not exactly the same as the partial pressure, the product of CO_2 mol fraction (x'_{CO_2}) times total gas pressure, as presented above (in Eq. 5.8). Because most readers are more familiar with the pCO_2, than fCO_2, we will use the former notation, noting that they are different.

Table 5.1 Acid-base reaction in seawater and examples of expression of their stoichiometric constants (Dickson et al. 2007). T is absolute temperature (K), S is salinity

Reaction	Equilibrium relations	Stoichiometric constant expression
$H_2O \Leftrightarrow H^+ + OH^-$	$K_w = [H^+][OH^-]$	$ln(K_w) = \dfrac{-13847.26}{T} + 148.9652 - 23.6521\,ln(T)$ $+ \left(\dfrac{118.67}{T} - 5.977 + 1.0495ln(T)\right)S^{0.5} - 0.01615S$
$CO_{2(g)} \Leftrightarrow H_2CO_3$	$[H_2CO_3] = K_H \times pCO_2$	$ln(K_H) = 93.4517\left(\dfrac{100}{T}\right) - 60.2409 + 23.3585\,ln\left(\dfrac{T}{100}\right)$ $+ S\left(0.023517 - 0.023656\left(\dfrac{T}{100}\right) + 0.0047036\left(\dfrac{T}{100}\right)^2\right)$
$H_2CO_3 \Leftrightarrow HCO_3^- + H^+$	$K_1 = \dfrac{[HCO_3^-][H^+]}{[H_2CO_3]}$	$ln(K_1) = \dfrac{-3633.86}{T} + 61.2172 - 9.67770\,ln(T)$ $+ 0.011555S - 0.0001152S^2$
$HCO_3^- \Leftrightarrow CO_3^{2-} + H^+$	$K_2 = \dfrac{[CO_3^{2-}][H^+]}{[HCO_3^-]}$	$ln(K_2) = \dfrac{-417.78}{T} - 25.9290 + 3.16967\,ln(T)$ $+ 0.01781S - 0.0001122S^2$
$B(OH)_3 + H_2O \Leftrightarrow B(OH)_4^- + H^+$	$K_B = \dfrac{[B(OH)_4^-][H^+]}{[B(OH)_3]}$	$ln(K_B) = \dfrac{-8966.90 - 2890.53S^{0.5} - 77.942S + 1.728S^{1.5} - 0.0996S^2}{T}$ $+ (148.0248 + 137.1942S^{0.5} + 1.62142S)$ $+ (-24.4344 - 25.085S^{0.5} - 0.2474S)\,ln(T) + 0.053105S^{0.5}T$

5.3 Analytical Parameters of the CO_2 System

Not all individual species of the CO_2 system (pCO_2, $[H_2CO_3]$, $[HCO_3^-]$, $[CO_3^{2-}]$, $[OH^-]$ and $[H^+]$) can be measured directly in seawater, and there is also no need to because they are interlinked via equilibrium, mass balance and charge conservation equations (Dickson 2011). There are five parameters that can be measured:

(A) The total concentration of dissolved inorganic carbon, often abbreviated as DIC, $\sum CO_2$ or C_T: **DIC** $= [H_2CO_3] + [HCO_3^-] + [CO_3^{2-}]$, is normally measured by acidifying the sample, stripping the evolved gas and measuring the total CO_2 content.

(B) The partial pressure (fugacity) of carbon dioxide (**pCO_2**) can be obtained by measuring the gas phase composition in equilibrium with the water and the use of Eq. (5.8: Henry's law).

(C) The **pH** can be measured directly using electrodes and by colorimetry using an indicator dye (Dickson 2011), which then provides data for the hydrogen concentration. However, there is also a large body of research based on the NIST scale (National Institute of Standards and Technology, formerly known as NBS scale), the total (pH_T) and the seawater (pH_{SWS}) scales. The latter two incorporate association of free protons with bisulphate, or bisulphate and fluoride, respectively, and are linked to the free scale as follows: $pH \approx pH_T + 0.11$ and $pH \approx pH_{SWS} + 0.12$.

(D) The carbonate ion concentration ($[CO_3^{2-}]$) via measurement of the ultraviolet absorbances of lead carbonate complexes in seawater.

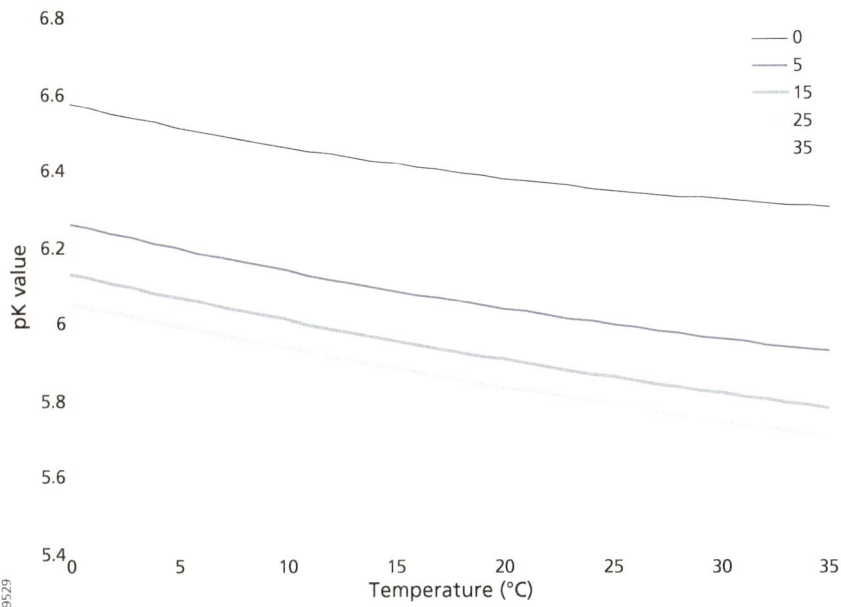

Fig. 5.2 The dependence of pK$_1$ on the temperature and salinity of water

(E)   The total or titration alkalinity (**TA**), which can be seen as the excess of proton acceptors over donors of a solution and is normally derived from an acidimetric titration. The total alkalinity of a solution is relatively easy to measure, and which has the pleasant characteristic that it behaves conservatively as two waters are mixed or if temperature or pressure changes, similar to DIC. A formal definition and detailed treatment follow below.

Let us revisit the simple CO$_2$–H$_2$O system from above with five species ([H$_2$CO$_3$], [HCO$_3^-$], [CO$_3^{2-}$], [OH$^-$] and [H$^+$]). Titration of this system with a strong acid results in a decrease in pH and transformation of carbonate to bicarbonate and bicarbonate to carbonic acid. Each bicarbonate ion can accept one proton, each carbonate ion consumes two protons before it is transformed into carbonic acid, and carbonic acid cannot accept protons. For this system, we can define H$_2$CO$_3$ as our reference (zero level of protons). Similarly, the hydroxide ions in solution will accept one proton, while any proton available just adds to the final proton concentration at the end of a titration. TA was formally defined by Dickson (1981) as the excess of proton acceptors ([HCO$_3^-$] + 2 [CO$_3^{2-}$] + [OH$^-$]) over proton donors [H$^+$] of the initial solution:

$$TA = \left[HCO_3^-\right] + 2\left[CO_3^{2-}\right] + \left[OH^-\right] - \left[H^+\right] \tag{5.11}$$

The reader might have noted that this TA definition has much similarity to the charge balance and proton balance Eqs. (5.9), (5.10a, b) for this simple CO_2–H_2O system. If fact by re-arranging the charge balance (5.9) we obtain the charge balance alkalinity or excess negative charge (ENC):

$$ENC = [HCO_3^-] + 2[CO_3^{2-}] + [OH^-] - [H^+] \tag{5.12}$$

and we see that this excess negative charge is equivalent to the titration alkalinity for this system. This excess negative charge introduced by Soetaert et al. (2007) is also known as the explicit conservative equation of total alkalinity (Zeebe and Wolf-Gladrow 2001; Wolf-Gladrow et al. 2007).

Although seawater is a much more complex system, because it contains many potential proton acceptors and donors, such as borate, phosphate and silicate species, the same approach can be used. Dickson (1981) formally defined total alkalinity of filtered seawater by the excess of proton acceptors over proton donors with respect to the proton condition at pH 4.5 (Dickson 1981), which then leads to the expression:

$$\begin{aligned} TA =& [HCO_3^-] + 2[CO_3^{2-}] + [B(OH)_4^-] + [OH^-] + [HPO_4^{2-}] + 2[PO_4^{3-}] + [H_3SiO_4^-] \\ &+ 2[H_2SiO_4^{2-}] + [NH_3] + [HS^-] + 2[S^{2-}] - [H^+] - [HF] - [HSO_4^-] \\ &- 2[H_2SO_4] - [H_3PO_4] - [HNO_2] - [HNO_3] \end{aligned}$$
$$\tag{5.13}$$

Following Wolf-Gladrow et al. (2007) and Soetaert et al. (2007), some additional proton acceptors ($H_2SiO_4^{2-}$, S^{2-}) and proton donors (HNO_2, HNO_3, H_2SO_4) have been added to the original definition of Dickson (1981). Most of these additional terms are close to zero in oxic seawater. Moreover, TA in marine waters is dominated by the carbonate species and simpler relations than (5.13) are often used. The carbonate alkalinity (CA) is defined as the alkalinity contribution of bicarbonate and carbonate species:

$$CA = [HCO_3^-] + 2[CO_3^{2-}] \tag{5.14}$$

and usually accounts for >95% of the alkalinity, with borate contributing another ∼3–4%. While the presence of these additional species hardly impacts measurement of TA by titration, it will affect the use of TA in calculations to obtain the sought-after distribution of bicarbonate and carbonate concentrations.

The excess negative charge concept, as introduced earlier (5.12) can also be applied to seawater (Soetaert et al. 2007; Wolf-Gladrow et al. 2007):

$$
\begin{aligned}
ENC =& \left[HCO_3^-\right] + 2\left[CO_3^{2-}\right] + \left[B(OH)_4^-\right] + \left[OH^-\right] + \left[H_2PO_4^-\right] + 2\left[HPO_4^{2-}\right] + 3\left[PO_4^{3-}\right] + \left[H_3SiO_4^-\right] \\
&+ 2\left[H_2SiO_4^{2-}\right] + \left[NO_3^-\right] + \left[NO_2^-\right] + \left[F^-\right] + \left[HS^-\right] + 2\left[S^{2-}\right] + \left[HSO_4^-\right] + 2\left[SO_4^{2-}\right] - \left[H^+\right] - \left[NH_4^+\right].
\end{aligned}
$$

$$(5.15)$$

ENC and TA are related via:

$$
TA = ENC + \sum NH_4 - \sum NO_3 - \sum NO_2 - \sum PO_4 - 2\sum SO_4 - \sum F
$$

$$(5.16)$$

where the various $\sum$-terms refer to total concentration of the respective acid-base pairs (e.g. $\sum NH_4 = [NH_4^+] + [NH_3]$). This difference between TA and ENC is caused by components for which the species used as zero proton levels are charged. In other words, DIC and $\sum B(OH)_3$ do not appear because they are uncharged at their zero proton level. This distinction between TA measured by titration (sensu Dickson) and ENC (charge balance alkalinity) is needed to understand the impact of biogeochemical processes on pH or pCO$_2$, in particular those involving ion-exchange or nutrient uptake (Zeebe and Wolf-Gladrow 2001; Soetaert et al. 2007; Wolf-Gladrow et al. 2007). For instance, nitrate or ammonium uptake by phytoplankton has a distinct effect on TA (decrease and increase, respectively; Brewer and Goldman 1976; Goldman and Brewer 1980) that cannot be inferred from the Dickson definition of TA (5.13), but is clear from Eqs. 5.15 and 5.16. The concept of potential alkalinity (Brewer et al. 1975) and its derivatives, such as TA*, are special cases of the ENC concept. The need to correct for nutrient cycling when using the alkalinity anomaly technique to quantify calcification is also made explicit when using Eq. 5.16.

5.4   Buffering

Natural waters, in particular seawater, contain multiple weak acids and their conjugate bases. Any addition or removal of an acid or base results in re-adjustment of the equilibrium distribution of acids and conjugated bases, with the consequence that the disturbance is attenuated (the principle of le Chatelier). Buffering can involve only dissolved components (homogenous) or also solid phases (heterogeneous). Heterogenous buffering in the ocean is called ocean carbonate compensation (Box 5.2). Homogenous buffering of waters is most often quantified by the buffering value (B), which is formally defined as the concentration of acid or base to be added to influence pH (Van Slycke 1922; Urbanksy and Schock 2000):

$$
B = \frac{dC_b}{dpH} = -\frac{dC_a}{dpH},
$$

$$(5.17)$$

Table 5.2 Buffer capacities expressed in proton concentrations (β) and pH (B)

Acids	$\beta = -\frac{dTA}{dH^+}$ (mol kg^{-1})	$B = \frac{dTA}{dpH}$ (mol kg^{-1})
H_2O	$-\frac{K_w}{[H^+]^2} - 1$	$\ln(10)\left(\frac{K_w}{[H^+]} + [H^+]\right)$
$B(OH)_3$	$-[B]_T \times \frac{K_B}{(K_B + [H^+])^2}$	$\ln(10)[B]_T \times \frac{K_B[H^+]}{(K_B + [H^+])^2}$
H_2CO_3	$-[CO_2]_T K_1 \times \frac{K_1K_2 + 4K_2[H^+] + [H^+]^2}{(K_1K_2 + K_1[H^+] + [H^+]^2)^2}$	$\ln(10)[CO_2]_T K_1[H^+] \times \frac{K_1K_2 + 4K_2[H^+] + [H^+]^2}{(K_1K_2 + K_1[H^+] + [H^+]^2)^2}$
H_3PO_4	$-[P]_T K_1 \times$ $\frac{4K_1K_2K_3[H^+] + K_1K_2[H^+]^2 + 4K_2[H^+]^3 + [H^+]^4 + K_1K_2^2K_3 + 9K_2K_3[H^+]^2}{(K_1K_2K_3 + K_1K_2[H^+] + K_1[H^+]^2 + [H^+]^3)^2}$	$\ln(10)[P]_T K_1[H^+] \times$ $\frac{4K_1K_2K_3[H^+] + K_1K_2[H^+]^2 + 4K_2[H^+]^3 + [H^+]^4 + K_1K_2^2K_3 + 9K_2K_3[H^+]^2}{(K_1K_2K_3 + K_1K_2[H^+] + K_1[H^+]^2 + [H^+]^3)^2}$

Notes
(1) K_1, K_2, K_3 are the first, second and third equilibrium constant for polyprotic acids
(2) The total buffer capacity is obtained summing all acid-base systems contributing (Hagens and Middelburg 2016)
(3) The inverse of B and ß are the sensitivity factors

where C_b and C_a are the concentration of added base and acid, respectively. The term buffer value as used here is identical to the terms buffer capacity, intensity and index in use by analytical chemists, geochemists and engineers. Buffer values are always positive because solutions resist changes according to the le Chatelier principle; consequently, a minus is needed when considering acid addition. The buffer value is a continuous non-linear function, as it is the inverse of the slope of the titration curve, i.e. a derivative (Stumm and Morgan 1996). The buffering value of seawater is expressed using the total alkalinity:

$$B = \frac{dTA}{dpH}. \tag{5.18}$$

For notational simplicity, we present buffer value as a derivative, while it is a partial derivative because all other concentrations are kept constant when determining this buffer value by titration (Hagens and Middelburg 2016). Van Slycke's original definition of buffer value was in pH, but in seawater it is instructive to express buffering value in terms of proton concentration (β):

$$\beta = -\frac{dTA}{dH^+}. \tag{5.19}$$

Buffer values B and β are linked via $B = -\ln(10)_* H^+ {}_* \beta$. Table 5.2 gives the relevant equations for B (in pH) and β (in proton concentrations).

Total alkalinity includes multiple acid-base systems, and the overall buffering value can be obtained by summing the contributions of all the acid-base systems involved (Urbanksy and Schock 2000). The buffer value of seawater is mainly determined by the borate and carbonate contributions, and we therefore limited our calculations to these (and of course the contribution of water). The buffer value of seawater (S = 35, T = 15 °C, DIC = 2000 µM) is a function of pH and has two distinct maxima at pH 6 and 8.9, near the pK_1 and pK_2 of the carbonate system in seawater and a minimum at pH 7.4 (approximately midway between these maxima; Fig. 5.3). Within the pH range of 7.5–8.5 the buffer value increases with pH and can be approximated by the relation:

$$B \cong \ln(10)\frac{K_2 DIC}{H^+}, \tag{5.20}$$

showing that the buffer value of seawater in its normal range primarily depends on the DIC concentration, the pH and pK_2 value. Borate contributes about 15–22% to seawater buffering in this range.

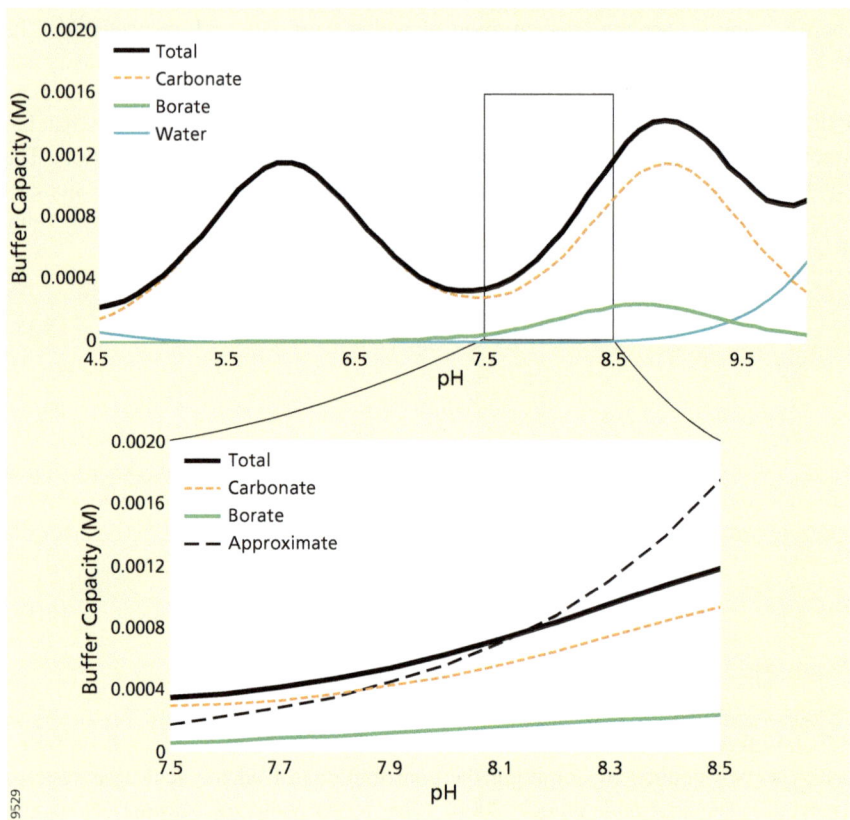

Fig. 5.3 Buffer capacity of seawater. Top panel: total capacity, carbonate, borate and water contributions. Lower panel: total capacity, carbonate and borate contributions and approximate buffer capacity (see text)

Another useful term related to the buffering is the sensitivity factor defined as the inverse of the buffer value expressed in terms of pH or protons (Soetaert et al. 2007; Egleston et al. 2010; Hagens and Middelburg 2016):

$$\frac{dpH}{dTA} = B^{-1} \tag{5.21}$$

$$-\frac{dH^+}{dTA} = \beta^{-1}. \tag{5.22}$$

These sensitivities and their use will be discussed below.

5.5 Carbonate Mineral Equilibria

In the above section, we only considered homogenous buffering, i.e. the re-equilibration among dissolved species. In natural systems, solid phases also contribute to buffering via dissolution and precipitation, and we call this heterogeneous buffering. Carbonate minerals play a dominant role in heterogeneous buffering because of their omnipresence, high solubility and high reactivity.

The precipitation and dissolution of calcium carbonate is represented by the reaction:

$$Ca^{2+} + CO_3^{2-} \Leftrightarrow CaCO_3 \qquad (5.23)$$

for which we can write the equilibrium relation:

$$K_{sp} = \left[Ca^{2+}\right]\left[CO_3^{2-}\right] \qquad (5.24)$$

where K_{sp} is the stoichiometric solubility product for calcium carbonate, which is a function of temperature, salinity, pressure and the calcium carbonate mineral. The solubility product is used to calculate the saturation state of seawater (Ω) with respect to the calcium carbonate mineral:

$$\Omega = \frac{[Ca^{2+}][CO_3^{2-}]}{K_{sp}}. \qquad (5.25)$$

If $\Omega = 1$, then the solution is in equilibrium with that mineral, if $\Omega > 1$, then the solution is supersaturated with respect to that mineral and if $\Omega < 1$, then the solution is undersaturated with respect to that mineral. Undersaturation and supersaturation promote dissolution and precipitation, respectively.

The two dominant calcium carbonate minerals in the ocean are calcite and aragonite, which differ in crystal structure, mineralogical properties (e.g. density) and solubility (Morse et al. 2007). The thermodynamic solubility products of calcite (pK = 8.48) and aragonite (pK = 8.30) differ by a factor 1.5, while their stoichiometric solubility products at salinity 35 and T = 25 °C are more than two orders of magnitude higher (pK = 6.36 and 6.18, respectively). The crystal structure of carbonates has consequences for the incorporation of other ions, with larger cations (e.g., Sr^{2+}, Ba^{2+}) preferably incorporated in the orthorhombic aragonite structure and smaller cations (e.g., Mg^{2+}, Mn^{2+}) fitting better in the hexagonal calcite structure. Dissolved magnesium concentrations in seawater are rather high (>5 times Ca^{2+}) and Mg incorporation is therefore substantial (up to $\sim 25\%$ for biogenic carbonates), with large consequences for solubility. The solubility of Mg-calcite with about 2% Mg is lower than that of pure calcite, while calcites with about 12–15% Mg substitution are five times more soluble than aragonite (Arvidson and Morse 2014).

5.6 Dissolved Inorganic Carbon Systematics

The conservative properties TA and DIC are often used as master variables to analyse the impact of environmental change on the marine CO_2 system, because the distribution of the species carbonic acid, bicarbonate and carbonate, and pCO_2 and pH depend on temperature and salinity. Figure 5.4 shows the temperature dependence of pH, pCO_2, bicarbonate and carbonate ion concentrations for seawater (TA = 2300 µM, DIC = 2000 µM). Increasing temperatures will cause an almost linear decrease in pH with a slope of –0.0155 pH/°C, while it will cause an exponential increase in pCO_2 values. Bicarbonate and carbonate ion concentrations will decrease and increase with increasing temperatures (slopes of about −0.55 and +0.42 µM/^{0}C, respectively).

There are also distinct dependencies with salinity (Fig. 5.4), but these only include the impact of salinity on stoichiometric equilibrium constants, not the larger effect of salinity on total alkalinity via charge balance. Increasing salinity results in a decline in pH and carbonate ion concentration and an increase in bicarbonate and pCO_2.

Relations among TA, DIC, pCO_2 and pH are often presented graphically in the form of TA versus DIC plots with isolines for pH or pCO_2 (Fig. 5.5) for seawater (T = 15, S = 35). The addition or removal of TA or DIC or any of its constituents on such a plot is a vector property (Deffeyes 1965). For instance, CO_2 addition will increase DIC concentration, but not affect TA and thus result in a horizontal vector pointing to the right; pCO_2 of the water will increase and the pH will decline. Addition of protons will decrease TA, but not impact DIC; the downward heading vertical vector indicates that pH will decline and pCO_2 will increase. The addition of HCO_3^- will impact both DIC and TA by one unit; the vector will have a slope of one and both pH and pCO_2 will slightly increase. The addition of CO_3^{2-} will increase DIC by one unit and TA by two units, and the resulting vector indicates a lowering in pCO_2 and increase in pH.

5.7 The Impact of Biogeochemical Processes

Most biological and physical exchange processes occur on longer time scales than the chemical equilibrium reactions, and any addition or removal of a compound due to biology and physics thus causes re-equilibrations following the le Chatelier principle; i.e. buffering occurs. The impact of biogeochemical processes on the dissolved inorganic carbon system can be analysed (graphically or numerically) using their impact on DIC and TA.

Gas exchange. Invasion of CO_2 from the atmosphere into seawater causes an increase in DIC, but does not change TA, with the consequence that pCO_2 increases and pH declines. The reverse process, an efflux of CO_2 from seawater to the atmosphere, results in pCO_2 decline and pH increase (Fig. 5.6). These CO_2 exchange processes may be initiated either by changing atmospheric mixing ratios of CO_2 or by cooling or warming of seawater.

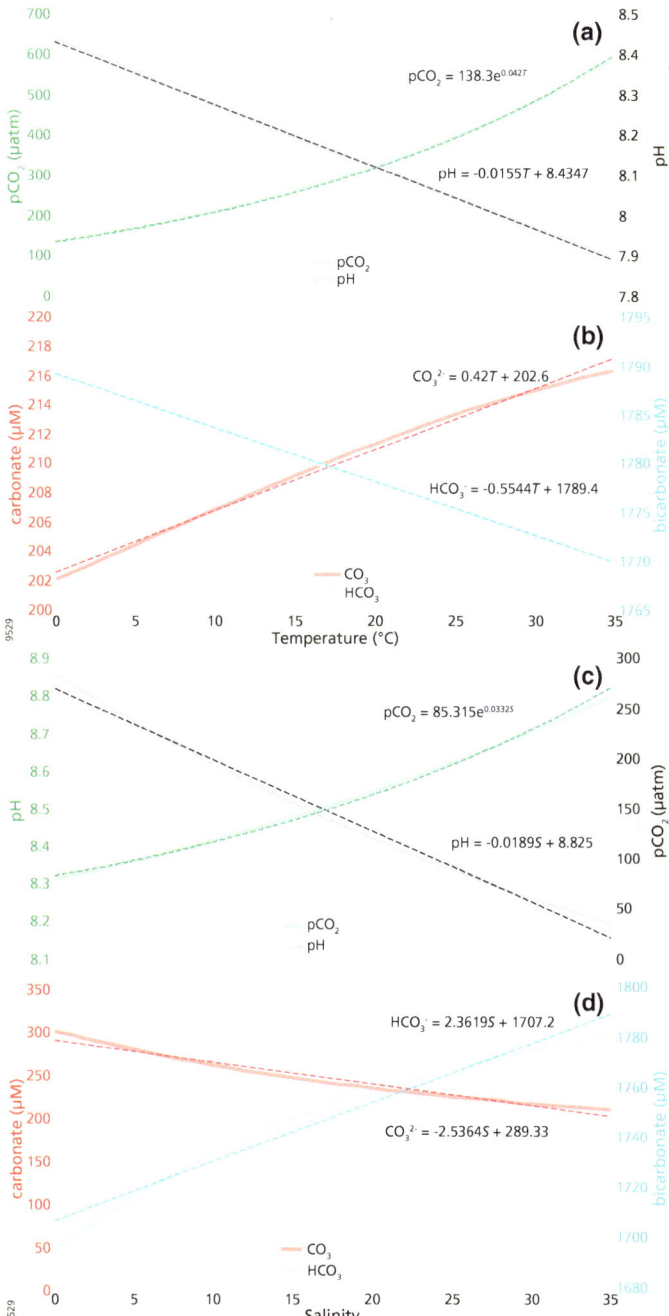

Fig. 5.4 **a** Dependence of seawater pCO_2 and pH on temperature; **b** Dependence of carbonate and bicarbonate ions on temperature; **c** Dependence of seawater pCO_2 and pH on salinity; **d** Dependence of seawater carbonate and bicarbonate ions on salinity. Dashed lines are regression equations

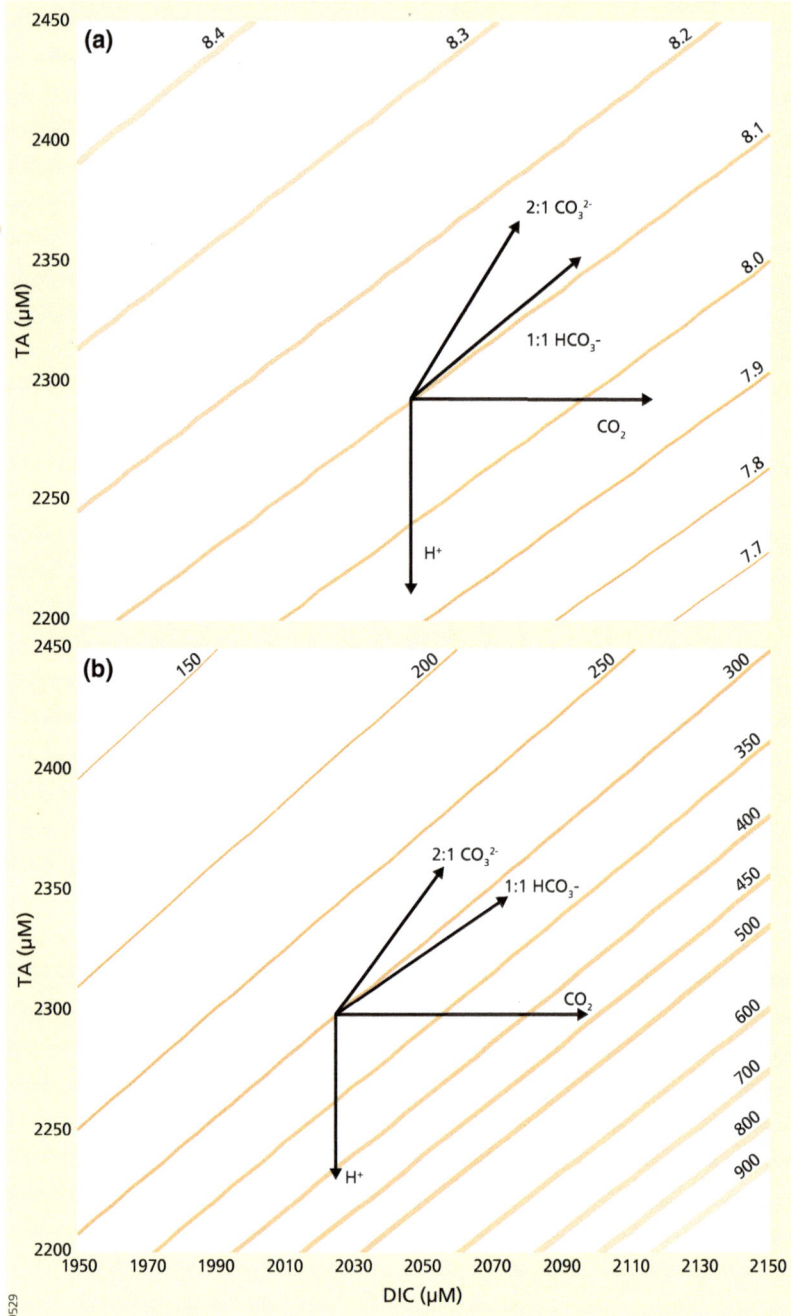

Fig. 5.5 TA-DIC plot with pH contours (**a**) and pCO$_2$ (µatm) contours (**b**). Vectors are given for addition of protons, carbon dioxide, bicarbonate and carbonate

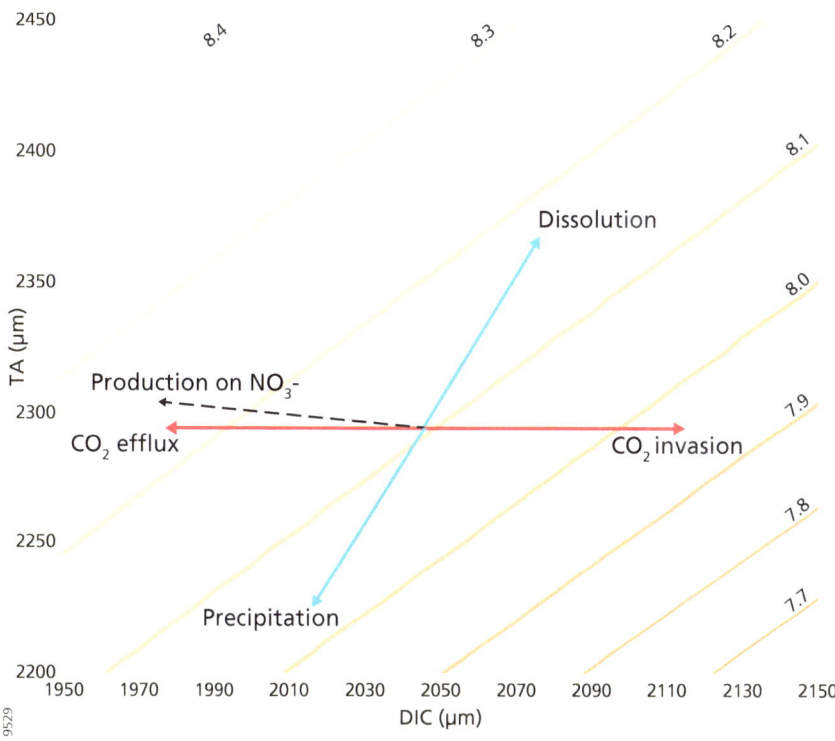

Fig. 5.6 Vector diagram on TA-DIC plot showing pH changes due CO_2 invasion and effluxes, carbonate mineral dissolution and precipitation, and primary production using nitrate as nitrogen source

Carbonate minerals. Precipitation of calcium carbonate via the reaction:

$$Ca^{2+} + 2\ HCO_3^- \Rightarrow CaCO_3 + CO_2 + H_2O \qquad (5.26)$$

removes one unit DIC from solution ($-2\ HCO_3^- + CO_2$) and two units of TA ($-2\ HCO_3^-$), with the result that pCO_2 increases and pH declines. Dissolution of calcium carbonates results in the release of one unit DIC and two units TA and thus a decline in pCO_2 and pH increase (Fig. 5.6).

The above examples directly involved one of the species of the dissolved inorganic carbon system, but many important biological processes do not directly involve any of these (e.g., oxidation-reduction reactions, such as nitrification or sulphide oxidation) or also involve other compounds besides the inorganic carbon species (e.g., primary production and organic matter degradation), and these do impact pH, pCO_2 and other parameters. To include these additional processes, we have to return to the relation between Excess Negative Charge (ENC) and Titration Alkalinity (TA).

Primary production involves the fixation of inorganic carbon and the assimilation of nitrogen and phosphorus to produce Redfield organic matter (see Chap. 6). Nitrogen assimilated in the form of either nitrate or ammonium has consequences for alkalinity because of the principle of nutrient-proton compensation mechanism (Brewer and Goldman 1976; Soetaert et al. 2007; Wolf-Gladrow et al. 2007). The rationale is as follows. Uptake of a nutrient in ionic form requires either uptake/release of a proton or hydroxide ion if internal compensation (e.g. NH_4^+ with PO_4^{3-}) is not occurring, otherwise the organism would be charged. Ammonium uptake results in a decrease in alkalinity because either a proton is released or an additional hydroxide ion is assimilated for charge compensation. By analogy, nitrate uptake increases alkalinity because either a proton is consumed or a hydroxide ion is released to maintain electroneutrality.

Accordingly, primary production based on ammonium can be represented as:

$$CO_2 + n\ NH_3 + p\ H_3PO_4 + H_2O \Rightarrow (CH_2O)(NH_3)_n(H_3PO_4)_p + O_2 \quad (5.27)$$

where n and p are the stoichiometric ratios (16/106 and 1/106 for Redfield ratios, respectively). This reaction lowers TA by n–p, i.e. 15/106 equivalents and DIC by one unit.

Similarly, primary production based on nitrate, i.e.,

$$\begin{aligned} CO_2 + n\ HNO_3 + p\ H_3PO_4 + (1+n)\ H_2O \\ \Rightarrow (CH_2O)(NH_3)_n(H_3PO_4)_p + (1+2n)\ O_2 \end{aligned} \quad (5.28)$$

increases TA with n + p, i.e. 17/106 equivalents and DIC by one unit (based on Redfield).

Note that these reactions have been presented in the form of CO_2, NH_3, HNO_3 and H_3PO_4 for convenience, but could have been written also in terms of bicarbonate, ammonium, phosphate and nitrate ions. Because of the electroneutrality condition and rapid equilibration, it does not matter which species is actually taken up. For instance, when CO_2 is taken up during photosynthesis, buffering will replenish CO_2 from the large bicarbonate pool, thereby consuming protons, i.e. increasing the pH. Alternatively, when bicarbonate is assimilated, electroneutrality maintenance would imply uptake of a proton or release of a hydroxide ion, the result being an increase in pH.

As another example, consider the anaerobic oxidation of methane coupled to sulphate reduction. We can write this equation either as

$$CH_4 + H_2SO_4 \Rightarrow H_2S + CO_2 + 2\ H_2O \quad (5.29a)$$

or

$$CH_4 + SO_4^{2-} \Rightarrow HS^- + HCO_3^- + H_2O \quad (5.29b)$$

Table 5.3 Biogeochemical processes and changes in excess negative charge and total alkalinity (Soetaert et al. 2007)

Process	Reaction	ΔENC	ΔTA
Aerobic mineralization	$(CH_2O)(NH_3)_n(H_3PO_4)_p + O_2 \Leftrightarrow CO_2 + n\,NH_3 + p\,H_3PO_4 + H_2O$	0	n–p
Denitrification	$(CH_2O)(NH_3)_n(H_3PO_4)_p + 0.8\,HNO_3 \Leftrightarrow CO_2 + n\,NH_3 + p\,H_3PO_4 + 0.4\,N_2 +$ $1.4\,H_2O$	0	0.8 + n– p
Mn-oxide reduction	$(CH_2O)(NH_3)_n(H_3PO_4)_p + 2\,MnO_2 + 4H^+ \Leftrightarrow CO_2 + n\,NH_3 + p\,H_3PO_4 +$ $2\,Mn^{2+} + 3H_2O$	4	n-p + 4
Fe-oxide reduction	$(CH_2O)(NH_3)_n(H_3PO_4)_p + 2\,Fe_2O_3 + 8H^+ \Leftrightarrow CO_2 + n\,NH_3 + p\,H_3PO_4 +$ $4\,Fe^{2+} + 5H_2O$	8	n-p + 8
Sulfate reduction	$(CH_2O)(NH_3)_n(H_3PO_4)_p + 0.5\,H_2SO_4 \Leftrightarrow CO_2 + n\,NH_3 + p\,H_3PO_4 +$ $0.5\,H_2S + H_2O$	0	n-p + 1
Fermentation	$(CH_2O)(NH_3)_n(H_3PO_4)_p \Leftrightarrow 0.5\,CO_2 + n\,NH_3 + p\,H_3PO_4 + 0.5\,CH_4 + H_2O$	0	n–p
Anaerobic oxidation of methane	$CH_4 + H_2SO_4 \Leftrightarrow CO_2 + H_2S + 2\,H_2O$	0	2
Calcite precipitation	$Ca^{2+} + CO_3^{2-} \Leftrightarrow CaCO_3$	–2	–2
Primary production (nitrate)	$CO_2 + n\,HNO_3 + p\,H_3PO_4 + (1 + n)\,H_2O \Leftrightarrow (CH_2O)(NH_3)_n(H_3PO_4)_p +$ $(1 + 2n)\,O_2$	0	p + n
Primary production (ammonium)	$CO_2 + n\,NH_3 + p\,H_3PO_4 + H_2O \Leftrightarrow (CH_2O)(NH_3)_n(H_3PO_4)_p + O_2$	0	p–n

n = N/C ratio of organic matter and p = P/C ratio of organic matter

Irrespective of the way formulated TA increases by two units and DIC by one unit. Soetaert et al. (2007) have worked out in detail the impact of biogeochemical processes on changes in ENC and TA (Table 5.3) and the related pH change. These pH changes are dependent on the actual pH of the system.

To illustrate the latter, we return to the precipitation of calcium carbonate. The precipitation of calcium carbonate can be presented as:

$$Ca^{2+} + CO_3^{2-} \Rightarrow CaCO_3, \tag{5.30a}$$

$$Ca^{2+} + 2\,HCO_3^- \Rightarrow CaCO_3 + CO_2 + H_2O, \tag{5.30b}$$

$$Ca^{2+} + HCO_3^- \Rightarrow CaCO_3 + H^+ \tag{5.30c}$$

$$Ca^{2+} + H_2CO_3 \Rightarrow CaCO_3 + 2H^+. \tag{5.30d}$$

While all these reactions are correct in terms of mass balance and stoichiometry: i.e. two units TA and one unit DIC are consumed for precipitation of one mole $CaCO_3$, the second one indicates production of CO_2 and the last two imply production of protons. This non-uniqueness is the result of re-equilibration reactions, and these have to be taken into account when presenting calcium carbonate precipitation or dissolution in a single equation. Evidently, at pH values $>pK_2$, the carbonate ion is the most important species (Fig. 5.1) and reaction (5.30a) will be the dominant reaction. This is the equation most often used in laboratory studies performed at high pH and no protons or carbon dioxide are net generated. Most field studies employ reaction scheme (5.30b) and have shown that calcium carbonate precipitation represents a source of carbon dioxide, but not one mole carbon dioxide for one mole of calcium carbonate because of buffering. Frankignoulle et al. (1994) presented the factor ψ which expresses the amount of carbon dioxide generated per unit calcium carbonate precipitated:

$$Ca^{2+} + (1 + \psi) \, HCO_3^- \Rightarrow CaCO_3 + \psi CO_2 + (1 - \psi) \, H^+. \qquad (5.31)$$

The factor ψ basically merges reactions (5.30b, c) above and can be calculated analytically and has a value of 0.6–0.7 for most marine waters. The result that pCO_2 increases and pH declines is consistent with the vector analysis presented above (Fig. 5.5).

An alternative generic treatment has been presented by Hofman et al. (2010).

$$Ca^{2+} + a_0 H_2 CO_3 + a_1 HCO_3^- + a_2 CO_3^{2-} \Rightarrow CaCO_3 + (2a_0 + a_1) \, H^+ \qquad (5.32)$$

where $\alpha_0 = H_2CO_3/DIC$, $\alpha_1 = HCO3^-/DIC$, $\alpha_2 = CO_3^{2-}/DIC$ and $\alpha_0 + \alpha_1 + \alpha_2 = 1$. The number of protons released by calcite precipitation is given by $2\alpha_0 + \alpha_1$, i.e. two times the carbonic acid + bicarbonate contribution to the DIC pool and varies from 2 at low pH to zero at high pH values (Fig. 5.7, red dashed line). However, the protons generated during calcite precipitation will be buffered, and the actual increase in proton concentration is given by $-\frac{2\alpha_0 + \alpha_1}{\beta}$ (Hofman et al. 2010; red solid line) where β is the buffer value in terms of protons introduced earlier (Eq. 5.22). The released protons will decrease the pH of the solution (blue dashed line). A similar analysis can be made for calcite dissolution: two protons are consumed during calcite dissolution at low pH, the net proton change is less because of buffering and the pH increase shows a non-linear response with two maxima (blue solid line).

Using the information in Table 5.3 and the approach of Soetaert et al. (2007), it is possible to calculate the impact of any process on pH and how it varies with pH. The pH dependence is the product of the change in negative charge (i.e. number of protons involved) and the sensitivity factor (Eq. 5.21). Figure 5.8 shows the pH dependence of the sensitivity factor with distinct maxima at pH 4.3, 7.3 and 9.9. Aerobic mineralization releases carbon dioxide, ammonium and phosphate. Carbon dioxide production decreases the pH for pH $>$ $\sim$5 (Fig. 5.9), in particular at pH 9.9 and 7.3 because of poor buffering, but pH increases for pH $<$ $\sim$5 because of ammonium release. The release of carbon dioxide does not impact the negative charge (or proton balance) at low pH because all inorganic carbon is already present as carbon dioxide. Denitrification lowers pH for pH $>$ $\sim$7, while it increases pH for pH $<$ $\sim$7 because of nitrate consumption and ammonium release, both impacting

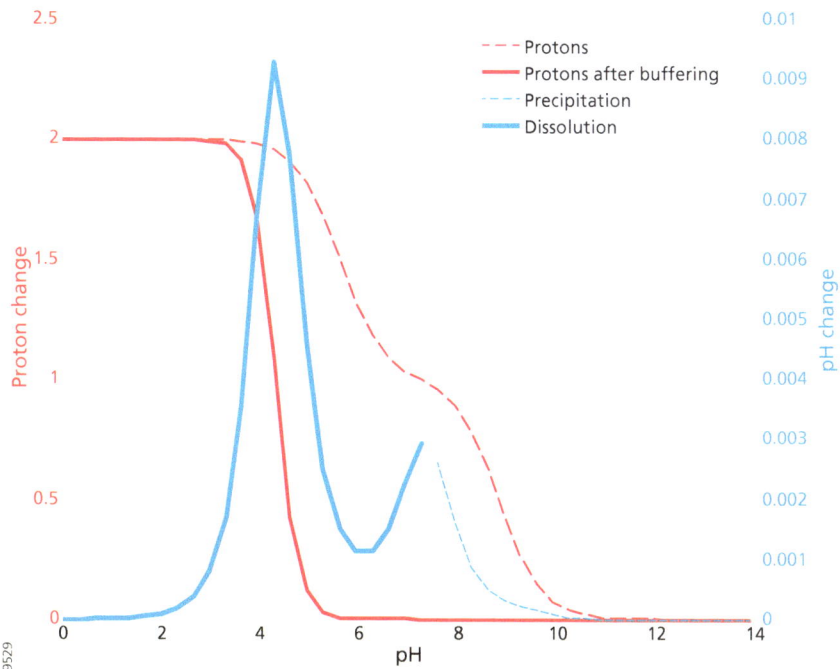

Fig. 5.7 Calcium carbonate precipitation/dissolution, proton and pH changes as function of pH. The number of proton generated per mole calcite formed/dissolved before buffering (red dashed line) and after buffering (red solid line). The decline in pH for precipitation of one µmole of calcite (dashed blue line) and increase in pH for dissolution of one µmole calcite (solid blue line). There is a break at the pH where calcite starts to dissolve or precipitate

the negative excess charge and TA (Eq. 5.16), in particular at pH 4.3 because of the high sensitivity factor (Fig. 5.8). Manganese and iron oxide reduction always increase pH with the relative pH increase primarily depending on the sensitivity factor (Fig. 5.9).

Box 5.1: Ocean acidification

Atmospheric carbon dioxide concentrations steadily increased during the last century because of the use of carbon-based energy resources, changes in land use, and lime production. Part of this additional carbon dioxide remains in the atmosphere, i.e. an airborne fraction of about 45%, and is the main driver of global warming. The other part ends up in the ocean or terrestrial biosphere. The ocean uptake of anthropogenic carbon accounts for 25–30%, but this service to humankind comes at a price: acid-base equilibria in the ocean have shifted. DIC and bicarbonate concentrations increased (carbonation), while carbonate ions and pH declined (ocean acidification). Ocean acidification, or

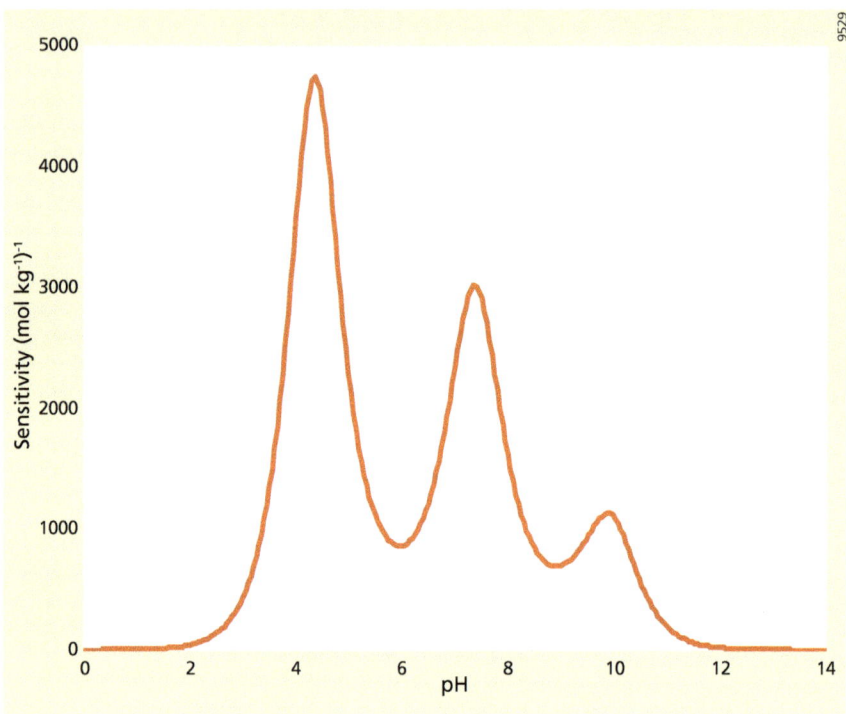

Fig. 5.8 The sensitivity factor $(\text{mol kg}^{-1})^{-1}$ as a function of the pH

the "other CO_2 problem", has consequences for the chemistry, biology and geology of the ocean.

Figure 5.10 shows the dependence of pH and DIC on atmospheric pCO_2 for a constant alkalinity ocean (2300 μM, T = 15, S = 35). Dissolved inorganic carbon concentrations increase almost linear with a slope of 0.513 μM/μatm pCO_2 (or 13.745 μM/μM dissolved CO_2), while pH declines almost linearly with a slope of –0.0011 pH/μatm pCO_2 (or –0.027 pH/μM dissolved CO_2). These thermodynamic predictions for declining pH and increasing DIC are fully consistent with observations in the ocean. These dependences of CO_2 and pH have received much attention and have been treated more formally using differential calculus (Sarmiento and Gruber 2006; Hagens and Middelburg 2016). The changes in pH can be split into multiple parts:

$$dpH = \left(\frac{\partial pH}{\partial T}\right)dT + \left(\frac{\partial pH}{\partial S}\right)dS + \left(\frac{\partial pH}{\partial TA}\right)dTA + \left(\frac{\partial pH}{\partial DIC}\right)dDIC + \cdots$$

$$(5.33)$$

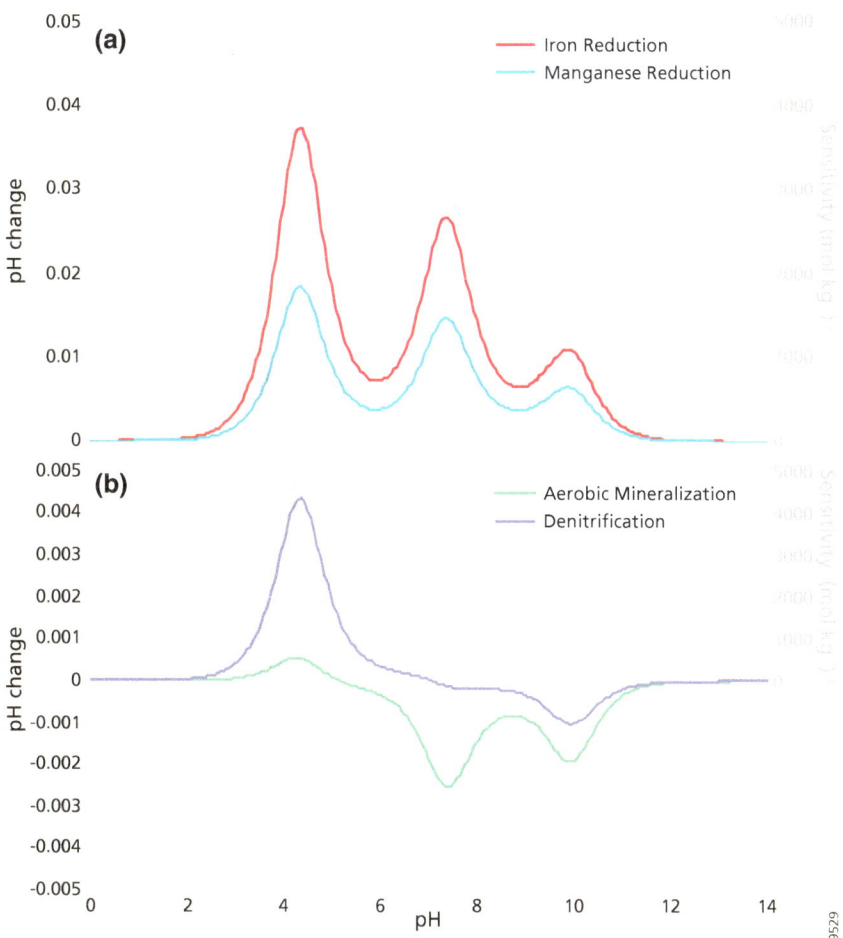

Fig. 5.9 **a** pH change due to iron-oxide and manganese oxide reduction as function of pH; **b** pH change due to aerobic mineralization and denitrification as a function of pH. The sensitivity factor (Fig. 5.8) is shown as reference (grey)

where $\left(\frac{\partial pH}{\partial x}\right)$ are partial derivatives, implying that these are, in principle, only valid if the other variables are kept constant, and they can be considered as a measure of the sensitivity of pH to a change in the respective environmental variable. For instance, the first partial derivative, $\left(\frac{\partial pH}{\partial T}\right)$ has a value of ~ -0.014 per degree (Hagens and Middelburg 2016), close to slope in Fig. 5.10 (–0.015 per degree). The third one, $\left(\frac{\partial pH}{\partial TA}\right)$, is the sensitivity factor (Eq. 5.21) shown in Fig. 5.8.

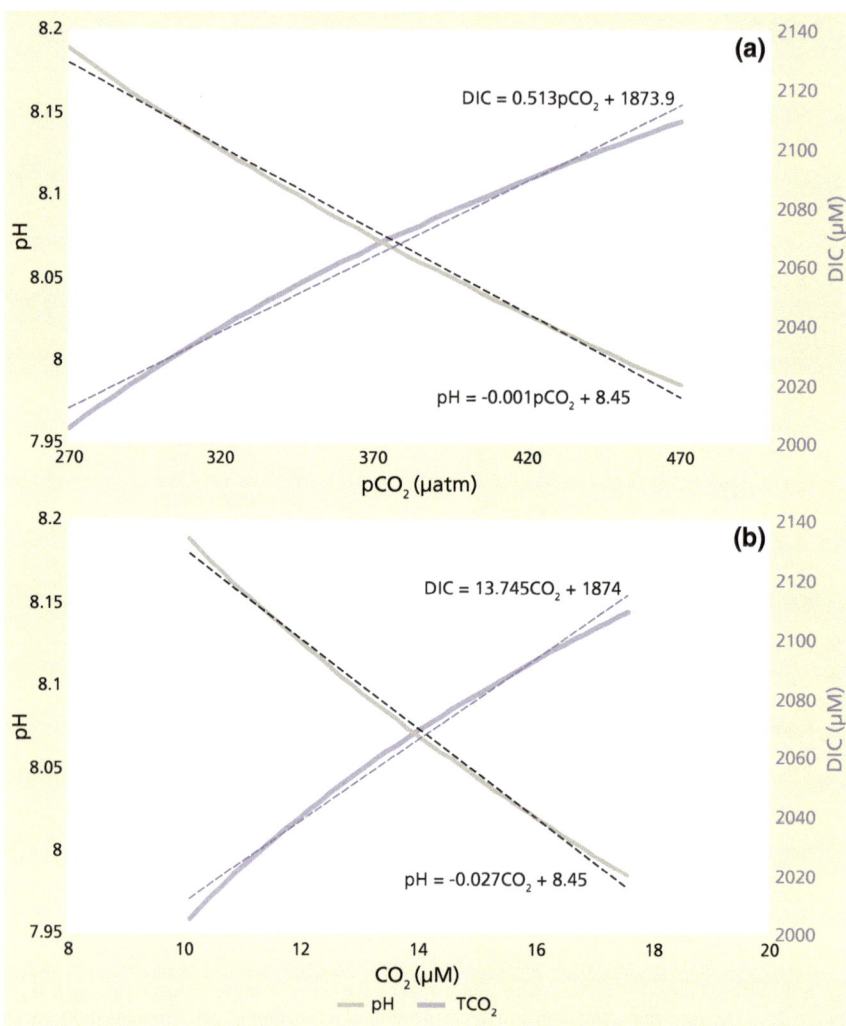

Fig. 5.10 pH and DIC dependence on pCO$_2$ (μatm) (**a**) and dissolved CO$_2$ (μM) (**b**). Regression lines are shown as dashed lines

Ocean acidification will lead to a decrease in buffer value (Fig. 5.3), and this is one of the reasons why the ocean is projected to take up less anthropogenic carbon in the future (besides changes in ocean physics and biology). Consequently, the ocean carbon dioxide system will be more sensitive to changes, and diurnal and seasonal changes in pH are projected to increase. Moreover, shifting chemical equilibria will have consequences for sound attenuation in the ocean.

Many organisms, autotrophs as well as heterotrophs, will be impacted by either the decline in pH and carbonate ions or the increase in dissolved inorganic carbon. The latter, carbonation, might stimulate primary producers, which are carbon limited at the moment. Most calcifying organisms are likely going to suffer from the decrease in carbonate ion availability, but calcification is under biological control and sometimes completed inside the organism, such that these responses are rather complex (Kroeker et al. 2013).

Box 5.2: Carbonate compensation dynamics

Buffering in the ocean not only occurs homogenously by re-arrangement of acids and conjugated bases, but also heterogeneously by re-adjustment of the balance between precipitation and dissolution of carbonate minerals. While homogenous buffering is fast (instantaneous equilibria), heterogeneous buffering has a larger capacity because of the large stock of carbonate minerals stored in marine sediments. This heterogeneous buffering is called carbonate compensation and biology plays a major role.

Carbonate minerals are predominantly formed biologically in the modern ocean, but there are a few exceptions, such as the formation of ooids in tropical systems and authigenic calcite and dolomite formation in sediments. However, even in these cases, biology plays indirectly a major role by governing the chemical composition of the fluids they formed in (e.g. carbonate formation induced by alkalinity production resulting from anaerobic oxidation of methane). Biological carbonate formation takes place in the water column by autotrophs (e.g., coccoliths) and heterotrophs (pteropods and foraminifera) and in the benthos by various organisms, autotrophs (e.g. coralline algae) and heterotrophs (e.g., corals, crustaceans, and molluscs). Biogenic carbonates can be aragonite, calcite or high-Mg calcites and combinations thereof.

Following death of calcifiers in benthic systems, the biogenic carbonate can either dissolve or accumulate at the seafloor and be buried. Carbonate produced in the surface layer of the open ocean can dissolve in the water column while particles settle, dissolve at the seafloor, or accumulate in sediments. Accumulation of biogenic carbonate is a prominent feature of ocean sediments; some sediments consists almost entirely of biogenic carbonate debris. Dissolution of carbonate is primarily driven by undersaturation, and organisms contribute to dissolution in a number of ways. One, many biogenic minerals have organic layers and microbial degradation of these layers exposes new surfaces to undersaturated solutions. Two, boring organisms (sponges, fungi) weaken the structure and texture of biogenic carbonate and consequently accelerate dissolution. Three, the metabolic activity of organisms has consequences for the saturation state of solution, e.g. carbon dioxide

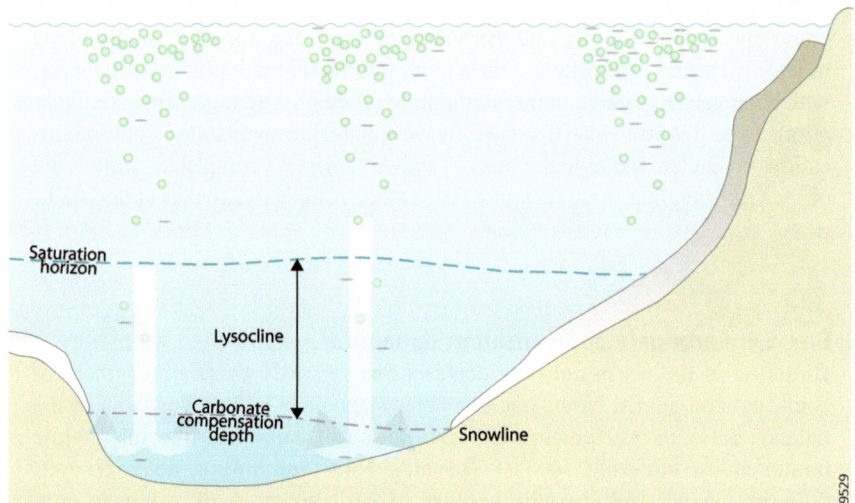

Fig. 5.11 Carbonate compensation concept. Biogenic $CaCO_3$ tests are produced in the photic zone of the oceans (green circles). Upon death, those tests escaping dissolution near the surface, settle along with clays materials. Above the saturation horizon, waters are supersaturated and $CaCO_3$ tests are largely preserved. Below the saturation, waters are undersaturated because of increasing solubility with depth and the release of CO_2 from organic matter decay and $CaCO_3$ will dissolve. Dissolution occurs primarily at the sediment surface as the sinking velocity of debris is rapid (broad white arrows). At the carbonate compensation depth, the rate of dissolution exactly matches rate of supply of $CaCO_3$ from above. At steady state this carbonate compensation depth is similar to the snowline; the first depth where carbonate poor sediments occur. The lysocline is the depth interval between the saturation and carbonate compensation depth (from Boudreau et al. 2018)

release during respiration and generation of strong acids during re-oxidation reactions (nitrification, sulphur oxidation) lowers the saturation state. The latter processes are important in coastal sediments that receive high organic carbon inputs and that are bioturbated or vegetated.

There are two carbonate compensation mechanisms operating in the ocean: chemical and biological compensation (Boudreau et al. 2018). Chemical compensation focuses on the dissolution or preservation of carbonate minerals at the ocean floor and implicitly assumes that net carbonate production remains constant. Carbonate particles settling in the ocean interior start to dissolve when water becomes thermodynamically undersaturated at depth z_{sat}. This carbonate saturation depth can be estimated from:

$$z_{sat} = z_{ref} ln\left(\frac{[Ca^{2+}][CO_3^{2-}]}{K_{sp}}\right) \tag{5.34}$$

where K_{sp} is the temperature, salinity and pressure depending stoichiometric solubility product, $[Ca^{2+}]$ and $[CO_3{}^{2-}]$ are the concentration of dissolved calcium and carbonate ions and z_{ref} is a scaling parameter (Boudreau et al. 2010). This undersaturation is partly due to increasing pressure and declining temperature (thermodynamics) and partly the consequence of a decrease in carbonate ion because of respiration of particulate organic matter (the biological pump). Below z_{sat} dissolution rates increase systematically with depth, and at a certain depth the dissolution rate balances the settling flux of carbonate (Fig. 5.11). This carbonate compensation depth (z_{CCD}) is governed by the following equation (Boudreau et al. 2010):

$$z_{CCD} \approx z_{ref} \ln \left(\frac{F_{car}[Ca^{2+}]}{K_{sp}A\beta_{mt}} + \frac{[Ca^{2+}][CO_3^{2-}]}{K_{sp}} \right) \qquad (5.35)$$

where F_{car} is the export flux of carbonate, A is the surface area of the seafloor and β_{mt} is the mass transfer of solutes across the diffusive boundary layer at the seafloor. The carbonate compensation depth (z_{CCD}) is always larger than the saturation horizon (z_{sat}) because of the first term with only positive parameters. The carbonate compensation depth is equal to the snowline, the depth at which carbonate disappear from sediments, under steady-state conditions. During periods of ocean acidification, bottom-waters will eventually obtain lower carbonate ions concentrations, and the saturation and carbonate compensation depths rise and dissolution of carbonate at the seafloor increases until a new balance between dissolution and export flux of carbonate has been reached. Conversely, during periods of alkalinisation, bottom water will eventually get higher carbonate ion concentrations (by advection), carbonate mineral dissolution decreases, resulting in a deepening of saturation and compensation depth till a new balance has been reached. This carbonate compensation mechanisms operates on a time scale of 100–10,000 years.

This chemical compensation mechanism is based on the assumption that ocean acidification has no impact on calcification and the export of calcium carbonate. However, many experiments have shown that calcification rates, and thus export of calcium carbonate from the surface ocean declines with saturation state of surface water. A decline in carbonate export (F_{car}) would lead to shallowing of the carbonate compensation depth on the very short time scale (<1 yr), but would cause additional deepening on the longer term (>10^4 yr) because calcification is an alkalinity sink. With less removal of carbonate ions in the surface waters, deep water will eventually become richer in carbonate and more carbonate minerals will survive dissolution at the seafloor (Boudreau et al. 2018).

References

Arvidson RS, Morse JW (2014) Formation and diagenesis of carbonate sediments. Treatise Geochem (2nd edn.) 9(3):61–101

Boudreau BP, Meysman F, Middelburg JJ (2010) Carbonate compensation dynamics. Geophys Res Lett. https://doi.org/10.1029/2009gl041847

Boudreau BP, Middelburg JJ, Luo Y (2018) The role of calcification in carbonate compensation. Nat Geosci. 11:894–900.

Brewer PG, Goldman JV (1976) Alkalinity changes generated by phytoplankton growth. Limnol Oceanog 21:108–117

Brewer PG, Wong GTF, Bacon MP, Spencer DW (1975) An oceanic calcium problem? Earth Planet Sci Lett 26:81–87

Butler JN (1982) Carbon dioxide equilibria and their applications. Addison-Wesley Publishing Company.

Deffeyes KS (1965) Carbonate equilibria: a graphic and algebraic approach. Limnol Oceanogr 10:412–426

Dickson AG (1981) An exact definition of total alkalinity and a procedure for the estimation of alkalinity and total inorganic carbon from titration data. Deep Sea Res Part A Oceanogr Res Pap 28:609–623

Dickson AG (2011) The carbon dioxide system in seawater: equilibrium chemistry and measurements. In: Riebesell U, Fabry VJ, Hansson L, Gattuso JP (eds) Guide to best practices for ocean acidification research and data reporting. Publications Office of the European Union, Luxembourg, pp 17–40

Dickson AG, Sabine CL, Christian JR (eds) (2007) Guide to best practices for ocean CO_2 Measurements. PICES Special Publication, p 3

Egleston ES, Sabine CL, Morel FMM (2010) Revelle revisited: buffer factors that quantify the response of ocean chemistry to changes in DIC and alkalinity. Glob Biogeochem Cycles 24: GB1002

Frankignoulle M, Canon C, Gattuso J-P (1994) Marine calcification as a source of carbon dioxide: positive feedback of increasing atmospheric CO_2. Limnol Oceanogr 39:458–462

Goldman JC, Brewer PG (1980) Effect of nitrogen source and growth rate on phytoplankton-mediated changes in alkalinity. Limnol Oceanogr 25:352–357

Hagens M, Middelburg JJ (2016) Generalised expressions for the response of pH to changes in ocean chemistry. Geochmica et Cosmochimica Acta 187:334–349

Hofmann AF, Middelburg JJ, Soetaert K, Wolf-Gladrow DA, Meysman FJR (2010) Proton cycling, buffering, and reaction stoichiometry in natural waters. Mar Chem 121:246–255

Kroeker KJ, Kordas RL, Crim R, Hendriks IE, Ramajo L, Singh GS, Duarte CM, Gattuso J-P (2013) Impacts of ocean acidification on marine organisms: quantifying sensitivities and interaction with warming. Glob Change Biol 19:1884–1896

Millero FJ (2007) The marine inorganic carbon cycle. Chem Rev 107:308–341

Morse JW, Arvidson RS, Luttge A (2007) Calcium carbonate formation and dissolution. Chem Rev 107:342–381

Sarmiento J, Gruber N (2006) Ocean biogeochemical dynamics. Princeton University Press, Princeton, p 526

Soetaert K, Hofmann AF, Middelburg JJ, Meysman FJR, Greenwood J (2007) The effect of biogeochemical processes on pH. Mar Chem 105:30–51

Stumm W, Morgan JW (1996) Aquatic chemistry, chemical equilibria and rates in natural waters, 3rd edn. Wiley, New York, 1022p

Urbansky ET, Schock MR (2000) Understanding, deriving, and computing buffer capacity. J Chem Educ 77:1640–1644

Van Slyke DD (1922) On the measurement of buffer values and on the relationship of buffer value to the dissociation constant of the buffer and the concentration and reaction of the buffer solution. J Biol Chem 52:525–570

Wolf-Gladrow DA, Zeebe RE, Klaas C, Kortzinger A, Dickson AG (2007) Total alkalinity: the explicit conservative expression and its application to biogeochemical processes. Mar Chem 106:287–300

Zeebe RE, Wolf-Gladrow DA (2001) CO_2 in seawater: equilibrium, kinetics, isotopes. Elsevier Ltd., Amsterdam

Organic Matter is more than CH₂O

6

In the previous chapters, we have presented organic carbon in isolation from the other elements and focused on the biological processes involved in the transformation from inorganic to organic carbon and vice versa. This approach neglects the fact that the element carbon is part of molecules. The thermodynamic stability and reactivity of, and interactions among, these molecules eventually determine the function of organic carbon compounds and govern the rates of organic matter production, transformation and consumption. Organic molecules contain not only C, H and O, but also N, P, S and multiple other elements, and often in certain ratios, e.g. one P atom in each DNA or RNA nucleotide. Organic molecules have functional groups and stereochemistry that determine their interactions with the environment.

Following a concise discussion of Redfield organic matter and non-Redfield organic matter, we focus on organic matter as food for organisms, the compositional consequences of preferential consumption and the consequences for the composition of organic matter preserved and buried in marine sediments.

6.1 Redfield Organic Matter

Chemical analysis of marine particulate organic matter, organisms and seawater by early oceanographers inspired Redfield (1934) to propose that marine plankton have relatively constrained atomic ratios of ~ 140 C to ~ 20 N and ~ 1 P (Fig. 6.1). This empirical Redfield ratio was originally based on the similarity of these ratios in seawater and in marine plankton, and these ratios have since been shown to apply to assimilation ratios during phytoplankton growth and regeneration ratios during marine organic matter degradation. Redfield further noted the implications for oxygen consumption and later on (Redfield 1958) posed the intriguing question whether the biological processes control the proportions of these elements in the water.

© The Author(s) 2019
J. J. Middelburg, *Marine Carbon Biogeochemistry*, SpringerBriefs in Earth
System Sciences, https://doi.org/10.1007/978-3-030-10822-9_6

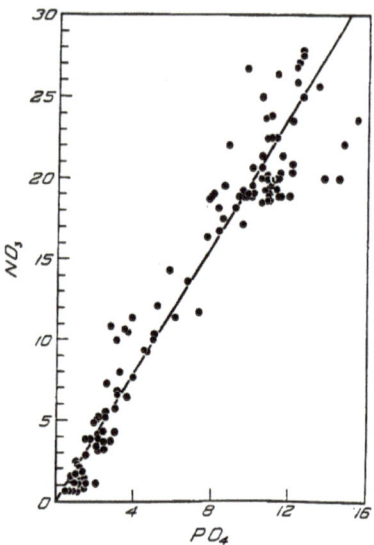

Fig. 6.1 Redfield's original correlation between phosphate and nitrate in seawater (Redfield 1934). The line has a slope of 20 rather than the modern value of 16 for ideal Redfield

The Redfield ratio as formalized by Redfield et al. (1963), at 106C:16 N:1P, has been validated by thousands of observations, and the overall ratios appear to be robust. Redfield ratios have shown useful in a wide range of topics, from nutrient limitation to biogeochemical modelling in the present and past ocean. They are used in reconstruction of anthropogenic carbon inventories and are in the equations underlying Earth System modelling (Sarmiento and Gruber 2006). The original Redfield ratio has been extended to include other elements and heterotrophic organisms (ecological stoichiometry) and has been related to cellular level and global scale processes (Falkowski 2000; Sterner and Elser 2002; Box 6.1). At the cellular level, nitrogen is primarily used for the construction of proteins and phosphorus for the synthesis of ribosomal ribonucleic acids (rRNA). Loladze and Elser (2011) have shown that protein and ribosome synthesis results in a protein: rRNA ratio of ∼3, which corresponds to an atomic N:P ratio of 16, as observed in the ocean for plankton. Moreover, they have shown that N and P limitation during cell growth results in N:P ratios below or above the Redfield ratio, respectively.

Redfield et al. (1963) presented the (canonical) overall reaction for marine organic matter production:

$$106\,CO_2 + 16\,HNO_3 + H_3PO_4 + 122\,H_2O \Rightarrow (CH_2O)_{106}(NH_3)_{16}H_3PO_4 + 138\,O_2$$

$$(6.1)$$

and consumption:

$$(CH_2O)_{106}(NH_3)_{16}H_3PO_4 + 138\,O_2 \Rightarrow 106\,CO_2 + 16\,HNO_3 + H_3PO_4 + 122\,H_2O$$

$$(6.2)$$

These RKR equations indicate that 138 units of oxygen are required for oxidation of one mole of Redfield organic matter: 106 oxygen molecules to convert CH_2O to CO_2 and 32 oxygen molecules to convert the NH_3 all the way to HNO_3. The O_2:C coefficient for aerobic respiration (including nitrification) is thus 1.3. Primary production based on nitrate (new production) or ammonium (recycled productions) has consequences for the quantity of oxygen released.

The problem with the traditional Redfield ratios as formulated in Eqs. 6.1 and 6.2 is that it simplifies marine organic matter into an elementary carbohydrate (CH_2O), whereas newly produced marine organic matter is better represented as a mixture of proteins ($\sim 65\%$), lipids ($\sim 19\%$), and carbohydrates ($\sim 16\%$) and other compounds (pigments and nucleic acids; Hedges et al. 2002; Bianchi and Canuel 2011). While simple carbohydrates are the first molecules formed during carbon fixation, the biosynthesis of proteins, lipids and polysaccharides results in the loss of hydrogen and oxygen. Hydrogen and oxygen loss results from dehydration, wherein CH_2O is transformed into CH_2 chains in proteins and lipids. Moreover, there is also hydrogen loss during the transformation of the NH_3-group in ideal Redfield organic matter to the NH-amide group in proteins. Estimates of marine organic matter based on mixing of proteins, lipid, carbohydrates and nucleic acids and using modern analytical tools (Anderson 1995; Hedges et al. 2002), arrive at average compositions of $C_{106}H_{175}O_{42}N_{16}P$ (Anderson 1995) and $C_{106}H_{177}O_{37}N_{16}PS_{0.4}$ (Hedges et al. 2002) rather than $C_{106}H_{260}O_{106}N_{16}P$ of the ideal Redfield ratio. This lower oxygen content has consequences for the O_2:C respiration coefficient. Equation 6.2 can be generalized to:

$$C_\alpha H_\beta O_\chi N_\delta P_\varepsilon + \gamma O_2 \Rightarrow \alpha CO_2 + \delta HNO_3 + \varepsilon\, H_3PO_4 + \lambda\, H_2O \qquad (6.3)$$

where $\gamma = \alpha + 0.25\ \beta - 0.5\chi + 1.25\delta + 1.25\ \varepsilon$ and $\lambda = -0.5\delta - 1.5\ \varepsilon + 0.5\ \beta$. Substitution of these revised marine organic matter compositions in Eq. 6.3 results in O_2:P ratios of 150 to 154 and an O_2:C respiration coefficient (γ/α) of ~ 1.43. The latter result is consistent with the O_2:C ratio of ~ 1.46 obtained from inverse analysis of water-column nutrient and oxygen changes along isopycnals (Anderson and Sarmiento 1994).

6.2 Non-redfield Organic Matter

The composition of marine particles is rather uniform because phytoplankton and bacteria (and their remains) dominate plankton, and these organisms are primarily made up of proteins (>50%) supplemented by lipids, carbohydrates, nucleic acids and pigments. Individual species may deviate from ideal or revised Redfield ratios, but over larger spatial and temporal scales these differences average out. The composition of seagrasses and, in particular, emergent saltmarsh vegetation and mangrove trees deviates from unicellular primary producers: higher C:N and C:P ratios, less proteins and nucleic acids and more structural carbohydrates and lignins.

Moreover, there is much more variability in the composition of benthic primary producers, particularly in macrophytes. The higher variability is due to tissue differentiation: compounds involved in photosynthesis and carbon fixation are primarily in the leaves and not in the roots. Carbohydrates for storage (e.g. starch) can be transferred from leaves to rhizomes for long-term storage. These marine macrophytes re-allocate compounds and essential nutrients over the growing season: e.g. the resorption of nitrogen from senescent leaves and transfer to active growing leaves. Structural carbohydrates and lignins are needed for strength to deal with currents, waves and wind. Lignins are cross-linked phenolic polymers that occupy the space between structural carbohydrates, such as cellulose in the cell wall of macrophytes and trees. Polysaccharides are permeable to water, while the more hydrophobic lignins are not. Crosslinking between lignins and celluloses thus not only strengthens the cell wall but also provides capability to transfer water efficiently from roots to the leaves, where evapotranspiration occurs. The lignin-carbohydrate provided strength and stiffness are needed for saltmarsh plants and trees to stand and for submerged vegetation to resist currents and waves. These polymers are relatively rich in carbon and poor in nitrogen and phosphorus; this is the reason why seagrass, salt-marsh plants and in particular mangroves have high C:N and C:P ratios. These polymeric substances are difficult to degrade, with the consequences that the palatability of seagrasses, marsh vegetation and mangroves is relatively low (see Fig. 3.10).

6.3 Organic Matter is Food

Organic matter, detritus as well as living biomass, is consumed because it is the primary source of energy and nutrition for heterotrophic consumers, i.e. it is food for animals and a substrate for heterotrophic microbes. Osmotrophic organisms either use dissolved organic matter directly or after extracellular (cell-attached or free) hydrolysis of polymeric substances into smaller units that can pass the cell membrane of these microbes. Animals ingest particulate organic matter, part of which is digested, and part is egested. The energy balance for animals is (Welch 1968):

$$I = G + R + E \tag{6.4}$$

where I is the ingestion of food, G is growth and reproduction of the animal, R is respiration and E is egestion and excretion (e.g. feces). Food ingested by animals is digested physically (e.g. chewing, grinding), chemically (by enzymes) and biologically (by microbes in the digestion organs) to make it amenable to enter tissues, where it can be used for growth and respiration (energy). This simple equation can be used to define assimilation (A = G + R), assimilation efficiency (A/I), gross growth efficiency (G/I) and net growth efficiency (G/A).

Moreover, during periods of sufficient food, consumers can internally store part of the food assimilated for reserves in the form of carbohydrates and lipids.

Assimilation efficiencies vary widely depending on the food quality and the consumer: carnivores generally have higher assimilation efficiencies than herbivores and detritivores. However, this is partly compensated by higher net growth efficiencies for the herbivores, with the result that gross growth efficiencies vary over a rather narrow range (15–35%, Welch 1968). Assimilation efficiencies vary between elements, between biochemical classes of compounds (carbohydrates, amino acids, lipids) and among molecules within a class of compounds. For instance, carbohydrates and amino acids enriched in intracellular materials have higher assimilation efficiencies than those in cell-wall materials because cell-walls are less digested (Cowie and Hedges 1996). The composition of the diet and the material assimilated often differs from the composition of the consumers; consequently, some compounds (e.g. carbohydrates) are more used for respiration, while other compounds (e.g. amino acids) may be preferentially directed towards growth and synthesis of new tissues.

Detailed feeding studies of lipids, carbohydrates and amino acids at the compound levels with marine animals have revealed that some compounds are preferentially consumed and used for respiration, while others are used to synthesize compounds de novo or by transformation of assimilated compounds. These patterns have been shown not only to be taxon specific, but also to have a microbial processing signature (Woulds et al. 2014). The latter is consistent with the microbiome concept, i.e. microbes within animals' digestion system are key to the functioning of the animal. Moreover, there are a number of compounds that are

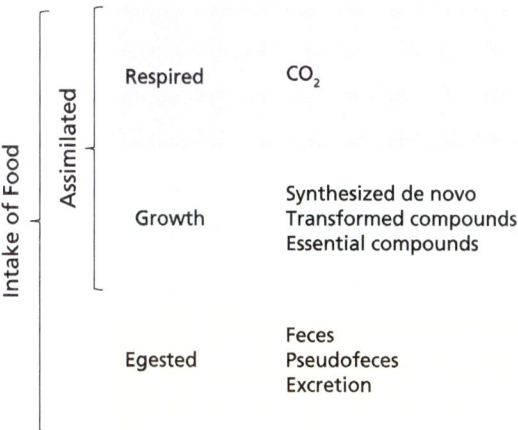

Fig. 6.2 Animal processing of organic matter. Food taken in is either digested or not and then egested. Digested organic matter is used for respiration or for growth that occurs primarily via synthesis of new compounds, but also via transformation or retention of compounds taken in. Moreover, some compounds can be produced de novo or transformed by microbes in the digestive system

considered essential because consumers cannot synthesize them and have to assimilate them from their diet or from the microbes in their digestive system. These include some poly-unsaturated fatty acids and amino acids, such as threonine, valine, leucine, isoleucine, phenylalanine and lysine. The combined effect of compound specific processing during digestion, the use of compounds for growth, storage or respiration, and the need to acquire essential compounds and elements results in changes in the organic matter remaining, excreted and produced by these consumers (Fig. 6.2).

6.4 Compositional Changes During Organic Matter Degradation

The above differences in assimilation efficiencies among various compounds and biochemical classes, and the preferential consumption of more labile components (Box 3.1; Fig. 3.9) have consequences for the composition of the organic matter remaining. Hedges, Wakeham, Lee and colleagues systematically studied the changes in the biochemical composition of organic matter during degradation from fresh phytoplankton, via sediment traps to surface and subsurface sediments (Wakeham et al. 1997). Amino acids (proteins), lipids and carbohydrates dominate the composition of phytoplankton and of detrital organic matter in shallow traps (Fig. 6.3). The composition changes rapidly upon degradation and organic matter in sediment traps below the surface mixed layer: amino acids decline from >60% to $\sim 25\%$ and lipids from ~ 10 to $\sim 2\%$, while the relative carbohydrate contribution remains similar because some carbohydrates are structural components. Moreover, the proportion of organic matter that cannot be characterized molecularly increases with progressive degradation from a few % in fresh phytodetritus to >50% in sediment trap organic matter and >70% in deep-sea sediment organic matter (Fig. 6.3).

The majority of organic matter in deep-sea sediment traps and sediments cannot be characterized using solution or gas-based chromatographic methods because of low organic solvent extraction and hydrolysis yields. Solid-phase NMR techniques have been applied as an alternative, and these studies not only confirmed the proportions of amino acid and carbohydrate based on solution-based techniques but also revealed that non-hydrolysable carbon-rich material (i.e. black carbon) and non-protein alkyl group made up most of the molecularly uncharacterizable organic matter (Fig. 6.4). Accordingly, the organic matter buried in sediment not only represents a small fraction of that produced (few %, Chap. 4), but it also differs significantly and systematically from phytodetritus (Fig. 6.3). This complicates the use of bulk organic matter properties as a proxy for the origin of the organic matter and for reconstruction of paleoenvironments (Middelburg 2018).

The susceptibility towards degradation varies systematically among biochemical classes: pigments > lipids = carbohydrate > amino acids > lignin > black carbon. Pigments, DNA, RNA and other cellular constituents are the most easily degradable

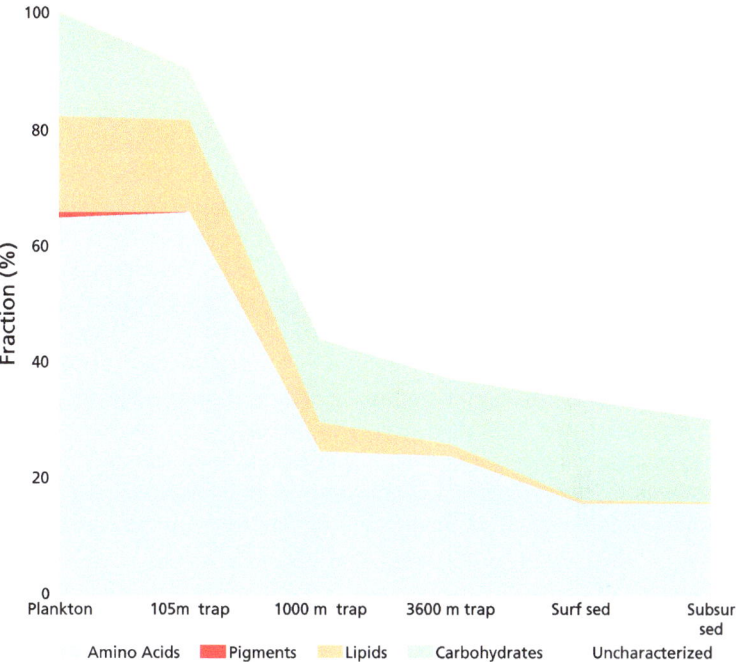

Fig. 6.3 Changes in the biochemical composition of organic matter during degradation from plankton, via sediment traps at different depths to surface and subsurface sediments in the Pacific Ocean (Wakeham et al. 1997). Uncharacterized fraction was calculated by difference: the uncharacterized fraction increases with progressive degradation

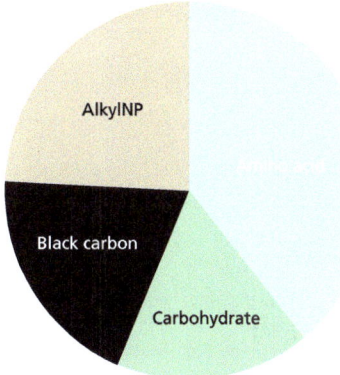

Fig. 6.4 Biochemical composition of marine sediments based on solid-phase NMR data (Gelinas et al. 2001). Four major compounds group were identified: amino acids, carbohydrates, black carbon and non-protein alkyl (AlkylNP)

compounds, followed by lipids, carbohydrate and amino acids (Fig. 6.3). While lignins are degraded in aerobic soils by fungi, these compounds are rather stable in the marine environments with the consequence that they have high preservation potential. Besides these differences among biochemical classes, there are also large differences among compounds within biochemical classes because of the inherent structural differences, as well as the macromolecular context. Structural carbohydrates are more stable than those involved in storage. Amino acids incorporated in large proteins are less available for consumers than those dissolved as free amino acids. Lipid reactivity towards degradation varies widely, as it depends on multiple factors, including the structure, stereochemistry, head-group, degree of saturation, and ether or ester linkages. While the relative reactivity of various compound classes and compounds within a biochemical class is systematic, absolute rates of degradation are largely context depending, i.e. identical compounds may have order of magnitude differences in kinetic parameters depending on the environment.

These systematic changes in organic matter composition due to degradation can be used to quantify the progress of degradation, or in other words, the degradation history. Often used degradation state parameters are intact to total pigment ratios,

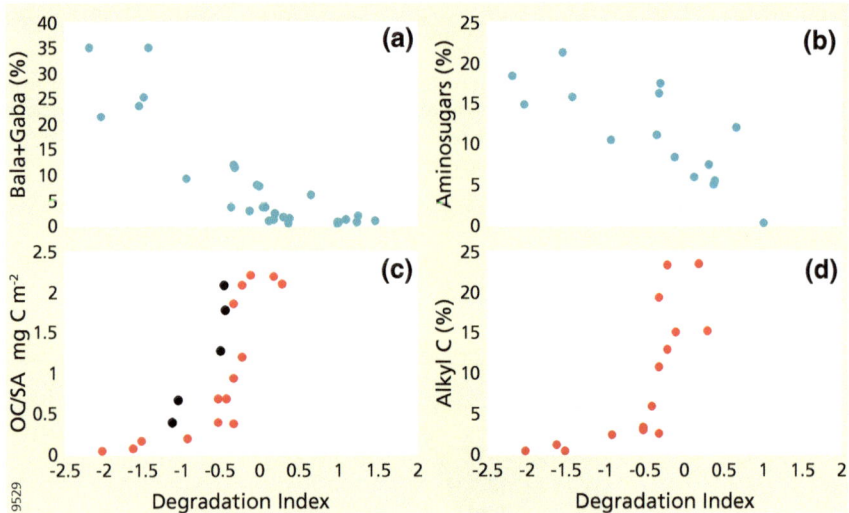

Fig. 6.5 Systematic changes in organic matter composition as degradation progresses. The degradation index of Dauwe and Middelburg (1998) is based on subtle changes in the amino acid composition. Fresh phytoplankton material has positive values, while extensively degraded sedimentary organic matter in deep-sea sediments has values <−1.5. Degradation causes relative accumulation of **a** non-protein amino acids Bala and Gaba and **b** aminosugars and a decrease in **c** the organic carbon content per sediment surface area (OC/SA) and **d** the fraction present as non-protein alkyl-C. Data are from Dauwe et al. (1999, blue), Gelinas et al. (2001, red) and Vandewiele et al. (2009, black)

pigment to carbon ratios, the contribution of amino acid carbon or nitrogen to the total carbon or nitrogen pool, the fraction of organic matter present as alkyl-C and the relative contribution of carbohydrates and amino acids to the total carbon pool. The most generic degradation state proxy is the amino acid based degradation index of Dauwe and Middelburg (1998). This degradation index is based on subtle changes in the amino acid composition of particulate organic matter. The resulting degradation index varies from +1.5 for fresh phytoplankton material to −2.5 for intensively degraded deep-sea sediment organic matter. The rationale is that the amino acid composition of phytoplankton is rather uniform and that changes measured in particulate organic matter can be attributed to mineralization processes. The degradation index correlates with multiple other organic matter degradation proxies, such as accumulation of microbial detritus (aminosugars, bacterial degradation products, D-amino acids derived from bacterial cell walls) and the decrease in organic carbon per unit surface area and contribution of alkyl-C (Fig. 6.5). The degradation index provides a continuous parameter to infer the organic matter degradation history, and one would expect a relation with the first-order rate constant for organic matter degradation shown in Fig. 3.9. Progressive degradation of organic matter should, according to the reactive continuum concept, lead to lower reactivity rate constant and to a lower degradation index. Figure 6.6 shows that the logarithm of the first-order rate constant and the degradation index of bulk organic matter are indeed correlated and that first-order rate constant can, in principle, be

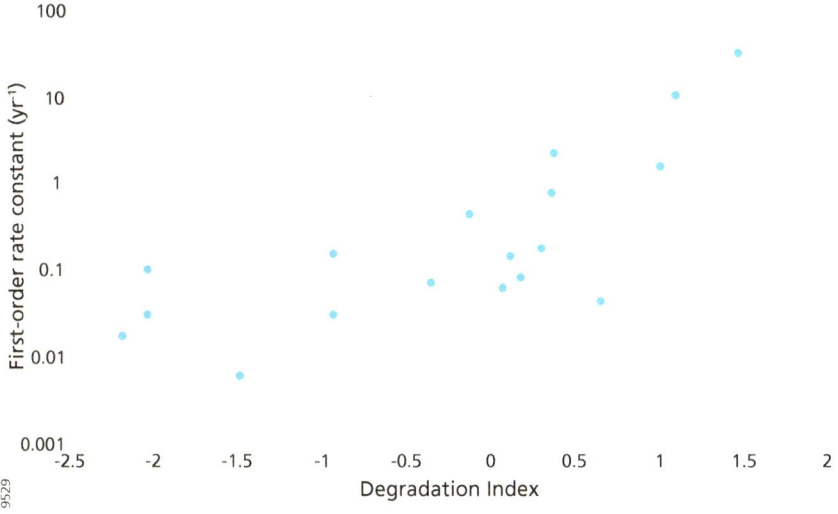

Fig. 6.6 Relation between the degradation index and the first-order reaction constant (data from Dauwe et al. 1999). Phytoplankton has a degradation index of 1 to 1.5 and a first-order rate constant of >10 yr^{-1}. Progressive degradation of organic matter lowers the degradation index and the reactivity constant

linked to organic matter composition. This does not imply that organic matter changes are the sole cause, but it does provide avenues to link kinetic biogeochemistry and organic geochemistry.

Box 6.1: Ecological stoichiometry

The traditional approach towards food-web and ecosystem functioning focuses on carbon, i.e. energy, flows and ignores the role of other elements and the biochemical makeup of organic matter. In other words, the traditional approach focuses only on quantity and largely ignores the quality of the organic matter. This is unfortunate, as all organisms need at least 20 elements or more and many organisms also have to acquire some essential compounds from the environment or their diet. Ecological stoichiometry explicitly deals with the flow of nutrients (primarily N, P, but also others) from the environments via primary producers to herbivores, detritivores and carnivores.

Inspired by the pioneering work of Redfield in the early 20th century, ecologists (Sterner and Elser 2002) have developed the theory of ecological stoichiometry in which homeostatis is a central concept. Homeostasis is the property of organisms to maintain a constant composition, despite living in a variable environment or feeding on a variable diet (Fig. 6.7). Some organisms lack homeostasis, and their cellular nutrient levels reflect that in the environment (you are what you eat), while strict homeostatic consumers maintain their composition. Heterotrophs are, in general, more homeostatic than autotrophs, but intermediate behavior has been reported. Phytoplankton has been shown to have high flexibility in term nutrients, but suboptimal nutrient contents come at the expense of performance (e.g. growth rate). Consumers living on resources that deviate from their tissue composition have multiple ways of elemental adjustment: food selection, feeding behavior, regulation of assimilation and metabolism (e.g. diverting towards respiration or excretion). Most herbivores are richer in N and P than primary producers; this stoichiometric mismatch is smaller in aquatic systems than in macrophyte systems in which carbon-rich lignin and structural cellulose are more abundant. Mismatches between the composition of a consumer and its resources has consequences for trophic transfer efficiencies and growth rates. The growth rate hypothesis involves a direct link between the growth rate, phosphorus content and ribosomes: i.e. high P content imply high RNA and thus high potential for growth. Ecological stoichiometry links the elements via compounds to the functioning of organisms and organism-scale processes to the global biogeochemical cycles.

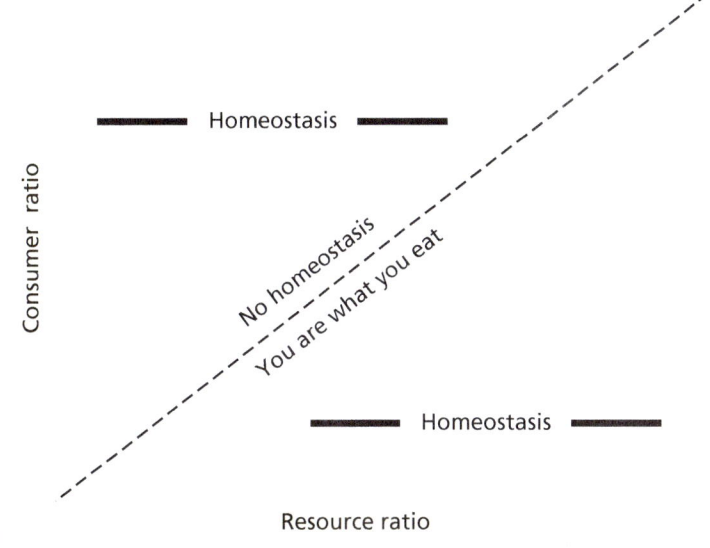

Fig. 6.7 Generalised stoichiometric pattern relating consumer (N:P, C:N, etc.) ratio to resource ratios. Organisms lacking homeostasis reflect the ratio of their resource (dashed 1:1 line), while strictly homeostatic organisms maintain their composition irrespective of the environment or their diet (horizontal line)

References

Anderson LA (1995) On the hydrogen and oxygen content of marine phytoplankton. Deep Sea Res Part I 42:1675–1680

Anderson LA, Sarmiento JL (1994) Redfield ratios of remineralization determined by nutrient data analysis. Glob Biogeochem Cycles 8:65–80

Bianchi TS, Canuel EA (2011) Chemical biomarkers in aquatic ecosystems. Princeton University Press

Cowie GL, Hedges JI (1996) Digestion and alteration of the biochemical constituents of a diatom (*Thalassiosira weissflogii*) ingested by an herbivorous zooplankton (*Calanus pacificus*). Limnol Oceanogr 41:581–594

Dauwe B, Middelburg JJ (1998) Amino acids and hexosamines as indicators of organic matter degradation state in North Sea sediments. Limnol Oceanogr 43:782–798

Dauwe B, Middelburg JJ, Herman PMJ, Heip CHR (1999) Linking diagenetic alteration of amino acids and bulk organic matter reactivity. Limnol Oceanogr 44:1809–1814

Falkowski PG (2000) Rationalizing elemental ratios in unicellular algae. J Phycol 36:3–6

Gelinas Y, Baldock JA, Hedges JI (2001) Carbon composition of immature organic matter from marine sediments: effect of oxygen exposure on oil generation potential. Science 294:145–148

Hedges JI, Baldock JA, Gélinas Y, Lee C, Peterson ML, Wakeham SG (2002) The biochemical and elemental compositions of marine plankton: a NMR perspective. Mar Chem 78:47–63

Loladze I, Elser JJ (2011) The origins of the Redfield nitrogen-to-phosphorus ratio are in a homoeostatic protein-to-rRNA ratio. Ecol Lett 14:244–250

Middelburg JJ (2018) Reviews and syntheses: to the bottom of carbon processing at the seafloor. Biogeosciences 5:413–427

Redfield AC (1934) On the proportions of derivatives in sea water and their relation to the composition of plankton. James Johnson memorial volume. University of Liverpool, Liverpool, UK, pp 176–192

Redfield AC (1958) The biological control of chemical factors in the environment. Am Sci 46:205–221

Redfield AC, Ketchum BH, Richards FA (1963) The influence of organisms on the composition of seawater. In: Hill MN (ed) The sea. Interscience, pp 26–77

Sarmiento J, Gruber N (2006) Ocean biogeochemical dynamics. Princeton University Press, pp 526

Sterner RW, Elser JJ (2002) Ecological stoichiometry: the biology of elements from molecules to the biosphere. Princeton University Press

Vandewiele S, Cowie G, Soetaert K, Middelburg JJ (2009) Amino acid biogeochemistry and organic matter degradation state accross the Pakistan margin oxygen minimum zone. Deep Sea Res II 56:376–392

Wakeham SG, Lee C, Hedges JI, Hernes PJ, Peterson ML (1997) Molecular indicators of diagenetic status in marine organic matter. Geochim Cosmochim Acta 61:5363–5369

Welch HE (1968) Relationships between assimilation efficiencies and growth efficiencies for aquatic consumers. Ecology 49:755–759

Woulds C, Middelburg JJ, Cowie GL (2014) Alteration of organic matter during infaunal polychaete gut passage and links to sediment organic geochemistry. Part II: fatty acids and aldoses. Geochimica Cosmochimica Acta 136:38–59